GET YOUR TECH TOGETHER

A Pedagogical Guide for Effective Instruction: Lessons from Instructional Leaders

Presented By: Jocelyn McDonald, Ph.D.

#TechItUp, LLC

Get Your Tech Together

Jocelyn McDonald, Ph.D.

Published by: #TechItUp, LLC

DrJocelynMcDonald.com

Print ISBN: 9798218081249

Printed in USA

Dedicated to

All Educators looking to make a positive impact in student lives and educational expereince.

TABLE OF CONTENTS

ACKNOWLEDGEMENTS

I would like to acknowledge the contributing authors of Get Your Tech Together: A Pedagogical Guide for Effective Instruction.

Thank you for sharing your knowledge, expertise, and experiences in education as we continue to create meaningful learning enviroments and make a positive impact in the lives of students.

Contributing Authors:

Dr. Travis Taylor

Dr. Affton Coleman

Eric Levels

Sherri Powell

Dr. Elena Silva-Leal

Arelys Barreto-Paz

Cicely Kelly Ward

Erin Carroll

Shirley Posey, M.S.

Dr. Kristina Tatiossian

VISIONARY AUTHOR

" Leverage technology as a vehicle to create meaningful learning experiences that mirror the digital world our students are brought up in and prepare them for the future they will contribute to. Lead. Inspire. Develop."

- Jocelyn "Dr. Mac" McDonald, Ph.D.

Introduction

by: Jocelyn McDonald, Ph.D.

Are you a K-12 educator looking to advance your instructional or leadership practice with technology? If so, this book was created just for you, the educator that may need more support or a better understanding of why and how technology enhances the instructional process and student experience.

Technology has been and will continue to be an ever-changing part of our lives, shaping our thoughts, daily interactions, and how we connect, communicate, and work collaboratively in a global society. Too often, we find ourselves comfortable with mediocre instructional practices because of a norm we have been accustomed to or feel is easier to implement; or even due to our personal ease of comfort levels and expectations. It is essential that we, as educators, do not lose sight that our purpose is to serve students and provide them with meaningful learning experiences to create a better future for themselves, our society, and the economy.

Now, I am a fan of the legendary singer and songwriter Prince. His song "1999" is my jam. But I often reference the song in education because I often find some educators' teaching and leadership practices are stuck back in time like it is still 1999 in 2022. For those who want to party like it's 1999, go for it. However, we have to stop teaching and leading like it's 1999. Think about how much has changed in the last two decades. We must prepare our students for their future, not our past. And sometimes, it may be uncomfortable to learn new skills and strategies to enhance instruction but ask yourself: Are you doing what's best for students? Are you creating a learning environment that will support students to become critical thinkers, problem solvers, adaptable learners, and global citizens who can equitably compete in a world with information overload at their fingertips? Again, this book was created for you, the educator who may need more support or a better understanding of why and how technology enhances the instructional process.

In the following chapters, you will find the voices of different instructional experts sharing instructional strategies, frameworks, and practices by taking you on a journey to how you can Get Your Tech Together.

Get Your Tech Together is organized into four sections:

Part 1: Foundations for Technology Integration

Part 2: Inclusive Instruction for All

Part 3: Science of Instruction to Increase Academic Performance

Part 4: Advancing Instruction with Innovation

You will find that this book is not a "how to use" a technology tool guide but a pedagogical guide for how to effectively integrate technology through the voices of different instructional leaders.

Contributing Authors

Dr. Travis Taylor is an instructional technology specialist in the Little Rock School District in Little Rock, Arkansas. Dr. Taylor has been in education for 28 years, 18 of which have been in educational technology. He provides professional development and training for teachers and administrators in current and emerging classroom technology and educational technology research.

Dr. Affton Coleman is a best-selling author, national presenter, elementary math specialist, curriculum writer, academic coach, educational consultant, and academic intervention implementation specialist. She has presented on technology integration at various conferences such as the Educator Rising National Conference, TCCA Conference, RTI Regional Conference, HAABSE, TABSE, & NABSE conferences, and other professional development sessions.

Eric Levels is the Executive Director of Digital Learning for Aldine ISD. He served as the Coordinator of Online Learning, facilitating Learning Management System implementation and training districtwide to over 3,000 staff members and 65,000 students. Levels have worked as creator and curator within the Professional Learning department facilitating training and development of the district through evaluation, education, and elevation of high-yield teaching strategies.

Sherri Powell's 23 years in education, passion to empower, dedication to innovation, and show-no-fear approach to technology makes her a positive change agent for education. She teaches and transforms technologically timid teachers and administrators into fearless educators who embrace the latest tools and resources to teach and engage their students. Sherri is also a Google Certified Trainer and Microsoft Innovative Expert and Trainer.

Dr. Elena Silva-Leal is in her twentieth year as an educator. Her background includes roles as a classroom teacher, department chair, teacher development specialist, and school administrator in private and public schools in urban areas. She specializes in support of turn-around campuses and has reached hundreds of teachers from across the state of Texas on inquiry-based instruction, scaffolds, digital integration, literacy routines, and data-driven instruction.

Arelys Barreto-Paz has served as a bilingual educator for many years and has always sought to integrate technology into instruction to support all learners across all content. Her experience in instructional technology extends from coaching teachers to managing a team of instructional technology coaches in a 1:1 technology middle school expansion program in a large urban school district serving over 200,000 students.

Cicely Kelly Ward has worked in public education over the past 22 years. She is committed to providing high-quality professional learning for campus leadership teams, teachers, and district staff to increase student achievement and data literacy, improving school-wide systems and classroom practice, and transforming leader and teacher efficacy. Her past roles have afforded her opportunities to coach and build data-driven instruction systems for schools and district staff.

Erin Carroll has served as an educator for twelve years. Her background includes roles as a youth counselor, social services case manager, classroom teacher, department chair, literacy specialist teacher leader, and literacy teacher development specialist in private and public schools in urban areas. In her latest role, Erin Carroll serves as an Academic Program Manager for Teacher Career Development Literacy Teacher leaders for the largest district in Texas.

Shirley Posey has been an educator and leader in education for over twelve years and is known for her expertise in evolving learners for the future. She has used her expertise as Director of STEM at Imhotep Institute Charter High School in Philadelphia, Adjunct Professor at Eastern University, and Johns Hopkins University CTY Instructor. Shirley consults and facilitates Professional Development for educators around the country on creating innovative STEM practices.

Kristina Tatiossian, Ph.D. ("Dr. Kris") is a scientist, entrepreneur, and consultant passionate about enabling the next generation of students to pursue STEAM careers. She has worked in top biotechnology companies developing novel therapeutics for the treatment of HIV, sickle cell disease, and more. Dr. Kris is the founder of CRISPR Classroom, an education technology company closing the gap between STEAM innovation and education.

“Strenghtening the foundation of your instructional practices with technology creates a path for discovery, practice, and reflection that leads to transformation.

- Jocelyn “Dr. Mac” McDonald, Ph.D.

PART 1

FOUNDATIONS FOR TECHNOLOGY INTEGRATION

DR. TRAVIS TAYLOR

"We continue having conversations suggesting how to recover from learning loss when we should be discussing how we elevate the educational system to address remote learning at a level of adequacy and confidence often seen in a face-to-face environment.

- Dr. Travis Taylor

Chapter 1

Teacher Presence in Online Learning

by: Dr. Travis Taylor

The Covid pandemic of 2020 impacted education in many ways, but none more than in the classroom. Literally, overnight, classrooms were transformed into spaces that many in the educational technology community talked about for years but were largely ignored. Now teachers, students, and parents were faced with ways of participating in education that many had never conceived of. Students and teachers were instantly distanced from each other, as were the students from their peers. Many educational leaders naively assumed that they could maintain a school in the same manner as before by using technology tools that they rarely embraced. Many thought their previous experience as leaders were adequate in guiding how schools could face this brand-new challenge, and many were stymied.

One of the main challenges to the classroom that was not observed was how distance would affect the teaching and learning experience. The phenomenon went unnoticed, partly because it was believed that technology would automatically facilitate many of the problems facing the new classroom experience. Still, nothing could have been further from the truth. Distance education, particularly with the use of technology, has been around for over a decade. Warnings from past eras were discussed then but seemed even more relevant now than ever. The forced distance between students and teachers in the classroom created a problem discussed in the 1990s. Wolcott (1996) described the phenomenon as "psychological distance" and how it diminishes the rapport within the classroom environment. At the time of the article, distance experienced by the classroom was generated from the evolution of distance learning in education. This distance, for the most part, was controlled or regulated in some capacity. Covid was an entirely new phenomenon. This was "forced distancing," where the consequences of violation were life-threatening, adding another stress level to teachers and students.

According to Wolcott (1996), rapport and interaction are damaged from long-term separation. Students and teachers may feel an increased sense of isolation which "contributes to a sense of apartness, that is, a failure to identify or affiliate with a group or an individual" (p. 25). Wolcott goes on to say, "In addition to feeling isolated, distant students may feel removed, alienated, or, at the extreme, disenfranchised" (p. 25). Although we are seeing declining waves of Covid infections, many in our society see Covid in the rear-view mirror. We still must address the after-effects. We saw many educators leave the profession during Covid; according to García and Weiss (2019) and Calhoun Williams (2022), this is only the beginning. The need for distance and remote education, as experienced in Covid, will be a part of a more extensive educational experience going forward. That being said, the need for classroom management in the virtual space is at the forefront of teaching and learning much as it was in face-to-face situations.

At the heart of the remote experience is the need to address the sense of community, rapport, and belonging. Much of the conversations around remote learning after the peak of Covid isolation focused on the term learning loss (Engzell, Frey, & Verhagen, 2021). This term referred to the amount of academic progress lost during the height of the pandemic. However, what has been implied with learning loss is misleading because it does not address the real need for learning in a remote environment. The learning loss was experienced because we did not know how to teach effectively in that space, not because remote instruction is inherently insufficient. We continue having conversations suggesting how to recover from learning loss when we should be discussing how we elevate the educational system to address remote learning at a level of adequacy and confidence often seen in a face-to-face environment. Simply put, we must evolve. We have seen from the data and the push for social-emotional learning that, as Wolcott expressed, extended remote educational experiences exacerbated the sense of being distanced.

Although I do not have all the answers to effectively teach in a remote setting, I can say that at the most basic level, establishing and maintaining reports and a sense of belonging is the foundation for classroom management in a remote environment. Teacher presence is essential because it is a constant within the classroom structure even when the class is not in session. Teacher presence is an ongoing function that facilitates the monitoring and adjusting of the social and cognitive dynamics of a classroom (Garrison, 1997; Richardson et al., 2012). Many of the aspects of educational and instructional technology are not mutually exclusive. They rely on or are affected by other technologies within the system. Levin and Schrum (2013) suggested that various aspects of the technology implementations must be maintained simultaneously. A failure or lack of attention in one could affect the entire system or implementation process. Therefore, this mirrors the same importance of teacher presence in online learning.

The relationship is dynamic because not all students enter the classroom with the exact emotional needs at equivalent levels. Teacher presence must regulate this dynamic. As we move our students into mastery of the lesson, there must be a transformation of ownership of the learning. This does not happen overnight, nor does it happen simultaneously for all students. The SAMR model (Puentedura, 2013) and the Technology Integration Matrix (TIM) framework (Welsh, Harmes, & Winkelman, 2011) support this progressive relationship where the teacher and student roles evolve over the lifetime of the classroom and lessons. This also holds true to Flock's (2020) and Garrison, Anderson, and Archer (2012) 's work on self-directed learning in the online environment. So, where does the teacher start with establishing teacher presence in the classroom? Here are some best practices to start as a foundation for creating rapport and establishing a safe and productive environment for learning.

Clarity

Give your students a fighting chance. What makes remote learning so difficult is reimagining the new classroom space. We often have the wrong idea about technology and how it should be used. We make assumptions about what should be easy and the relationship students have with technology. The digital native argument is a myth (Kirschner & Bruyckere, 2017; O'Neil, 2014; Scolari, 2019). I have seen students of various backgrounds respond differently to technology based on their socioeconomic status. Some students of low socioeconomic status rarely engage with computers and can be afraid to touch a computer for fear they may break something they feel they cannot afford. Be clear about dates, expectations, and deadlines. Understand how the students view the classroom and the setting and how they view their learning experience. Also, be sure to provide information and instruction based on how they view their learning experience and not how you assume they do.

Structure

One of the most essential advantages of online and remote learning is that the classroom can be "shaped," in a sense, or molded to fit the desired construct. We are not at the mercy of technology, but we get to determine what learning should look like with technology. Provide them with a safe space for learning by getting a vision and then designing a structure based on what your students need. As the teacher, you get to decide the setup and design of your class. Going back to the previous point, clarify this structure to the students. Describe the expectations of behavior, interaction, assignments, and performance in detail and with examples. This does not mean you cannot set high standards, but structure the class so this is evident.

Be Human

In the early work of adult learning theory, adult life experiences separated children from adults as learners because it was believed that students did not pay bills, make large purchases, or manage things such as insurance and mortgages. At that time, marginalized perceptions and student voices were not as relevant as adults (Knowles, 1979; Knowles, Holton, & Swanson, 2005). Looking at the changes in our society over the past generations, you will find that students are taking on more adult responsibilities than before. This can be a negative experience for students because they often experience situations that no child should experience, which have long-term consequences. The CDC refers to these situations as Adverse Childhood Experiences or ACE (Felitti et al., 2019). We know from research from the pandemic that many students did not want to be on camera for remote learning because they were ashamed of their living conditions being broadcast to their peers (Castelli & Sarvary, 2021).

You can set up the class by sharing things you like to do in your leisure time. By connecting students to experiences and life events outside of the classroom, teachers can be seen as just as human as they are. They can find out that their teachers can relate to their circumstances or the apprehension they may feel about being online or using technology.

Remote and online learning are here to remain, and we must have the courage to allow education to evolve to accommodate the mode of learning as mainstream. There are too many uncertainties in the world. The pandemic should have taught us that. But how to respond to the future of learning depends on how much we learned from this encounter. Remote learning is not changing teaching as much as it is changing the delivery of instruction and what it means to be a life-long learner.

5 TIPS TO SUPPORT IMPLEMENTATION

Clarity

Be clear about dates, expectations, and deadlines. Understand how the students view the classroom and the setting and how they view their learning experience.

Structure

We are not at the mercy of technology, but we get to determine what learning should look like with technology. Describe the expectations of behavior, interaction, assignments, and performance in detail and with examples.

Be Human

By connecting students to experiences and life events outside of the classroom, teachers can be seen as just as human as they are. You can set up the class by sharing things you like to do in your leisure time.

Don't Get Lost in the Features

Leverage technology to maximize learning for all students. Avoid trying to impress the students. It's not how much you can do. It's how much you can get done.

Continue to Learn

No one has all of the answers. Staying open to learning what you do not know increases your technology literacy, allowing you to have a larger amount of tools and resources available to you.

ABOUT DR. TRAVIS TAYLOR

Dr. Travis Taylor is an instructional technology specialist in the Little Rock School District in Little Rock, Arkansas. Dr. Taylor has been in education for 28 years, 18 of which have been in educational technology. He provides professional development and training for teachers and administrators in current and emerging classroom technology and educational technology research. Dr. Taylor specializes in multimedia applications and production, coaching teachers with methods to enhance student engagement and student voice. He also researches and makes recommendations for emerging technology implementation.

References

Calhoun Williams, K. J. (2022). The future implications of the faculty and staff shortage crisis. COSN 2022 Annual Conference. Nashville, Tennessee, USA.

Castelli, F. R., & Sarvary, M. A. (2021). Why students do not turn on their video cameras during online classes and an equitable and inclusive plan to encourage them to do so. Ecology and Evolution, 11, 3565. https://doi.org/10.1002/ece3.7123

Engzell, P., Frey, A., & Verhagen, M. D. (2021). Learning loss due to school closures during the COVID-19 pandemic. Proceedings of the National Academy of Sciences of the United States of America, 118(17). https://doi.org/10.1073/PNAS.2022376118

Felitti, V. J., Anda, R. F., Nordenberg, D., Williamson, D. F., Spitz, A. M., Edwards, V., ... Marks, J. S. (2019). Relationship of childhood abuse and household dysfunction to many of the leading causes of death in adults: The adverse childhood experiences (ACE) study. American Journal of Preventive Medicine, 56(6), 774–786. https://doi.org/10.1016/j.amepre.2019.04.001

Flock, H. S. (2020). Designing a community of inquiry in online courses. In International Review of Research in Open and Distributed Learning (Vol. 21). Retrieved from http://www.irrodl.org/index.php/irrodl/article/view/3985

García, E., & Weiss, E. (2019). The teacher shortage is real, large and growing, and worse than we thought The first report in "The Perfect Storm in the Teacher Labor Market" series Report •. In Economic Policy Institute. Retrieved from https://www.epi.org/publication/the-teacher-shortage-is-real-large-and-growing-and-worse-than-we-thought-the-first-report-in-the-perfect-storm-in-the-teacher-labor-market-series/

Garrison, R. (1997). Self-directed learning: Toward a comprehensive model. Adult Education Quarterly, 48, 18–33.

Garrison, R., Anderson, T., & Archer, W. (2012). Critical Inquiry in a Text-Based Environment. Computer Conferencing in Higher Education, 16(1), 6–12. https://doi.org/10.1016/j.sbspro.2011.12.092

Kirschner, P. A., & Bruyckere, P. (2017). The myths of the digital native and the multitasker. Teacher and Teacher Education, 67, 135–142. Retrieved from https://www.sciencedirect.com/science/article/pii/S0742051X16306692

Knowles, M. (1979). The adult learner: A neglected species. Educational Researcher, 8(3), 20. https://doi.org/10.2307/1174362

Knowles, M., Holton, E. F., & Swanson, R. A. (2005). A theory of adult learning: Andragogy. In The adult learner- The definitive classic in adult education and human resource development (6th ed., pp. 35–72). Retrieved from http://ebookcentral.proquest.com

Levin, B., & Schrum, L. (2013). Using systems thinking to leverage technology for school improvement: Lessons learned from award-winning secondary schools/districts. Journal of Research on Technology in Education, 46(1), 29–51. https://doi.org/10.1080/15391523.2013.10782612

O'Neil, M. (2014). Confronting the myth of the "digital native." The Chronicle of Higher Education, 21. Retrieved from https://internet.psych.wisc.edu/wp-content/uploads/532-Master/532-UnitPages/Unit-05/O'Neil_DigitalNatives_Chronicle_2014.pdf

Puentedura, R. (2013). The SAMR model: Six exemplars. Retrieved July 24, 2022, from hippasus.com website: http://www.hippasus.com/rrpweblog/archives/2012/08/14/SAMR_SixExemplars.pdf

Richardson, J. C., Arbaugh, J. Ben, Cleveland-Innes, M., Ice, P., Swan, K. P., & Garrison, D. R. (2012). Using the community of inquiry framework to inform effective instructional design. In L. Moller & J. B. Huett (Eds.), The Next Generation of Distance Education: Unconstrained Learning (Vol. 9781461417). https://doi.org/10.1007/978-1-4614-1785-9_7

Scolari, C. (2019). Beyond the myth of the "digital native" Adolescents, collaborative cultures and transmedia skills. Nordic Journal of Digital Literacy, 14, 164–174. https://doi.org/10.18261/issn.1891-943x-2019-03-04-06

Welsh, J., Harmes, J., & Winkelman, R. (2011). Florida's technology integration matrix. Principal Leadership, 69–71. Retrieved from https://www.setda.org/wp-content/uploads/2013/12/PLOct11_techtips.pdf

Wolcott, L. L. (1996). Distant, but not distanced: A learner-centered approach to distance education. https://link.springer.com/content/pdf/10.1007/BF02818902.pdf

DR. AFFTON COLEMAN

"SAMR is not just a guide to integrating technology. It's a guide to integrating ourselves into the digital age. It's a formula to revolutionize the learning environment for all learning types."

- Dr. Affton Coleman

Chapter 2

Revolutionize Instruction with SAMR

by: Dr. Affton Coleman

Have you ever felt stuck trying to keep up with the digital technology changes? No worries, you are not alone. Educators worldwide understand that technology is not going anywhere but do not know where to start when it comes to integrating technology into their daily or educational lives. In response to this is the SAMR model. It is a framework that assists educators, like yourself, with a framework to better understand and integrate technology.

Why SAMR?

SAMR is an acronym representing four levels of technology use: Substitution, Augmentation, Modification, and Redefinition. The model was founded on Ruben Puentedura's (2013) [1] research regarding what types of technology have the most significant effect on student learning. More specifically, SAMR supports using technology to engage

1 Puentedura (2013). SAMR: A contextualized introduction. Presentation given October 25, 2013 at Alberta Charter Schools Conference. Retrieved from http://www.hippasus.com/rrpweblog.

and drive student learning (Puentedura, 2013).[2] It became an essential influencer during the 2020 national pandemic. Educators and students took a massive setback in teaching and learning because they had to figure out how to teach in a virtual environment. And even more so for teachers who had never used or minimally used devices in their classrooms with educational lessons and technological applications to aid in delivering instruction. The pandemic uncovered various proficiency levels of teachers and administrators. Teachers began to show how they felt about technology integration and where they personally ranked in that manner after being quickly immersed into teaching online versus traditional classroom practices with students and partially integrating the technology during instruction. However, SAMR is a framework that can support and provide a systematic method incorporating technology.

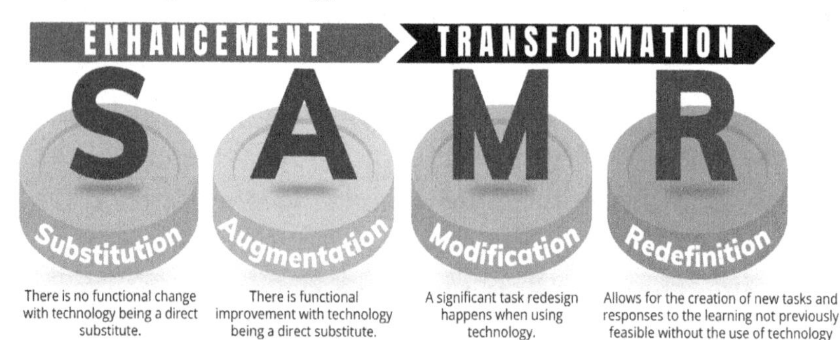

Understanding the Model

Using SAMR as an integration process can occur anywhere in the educational learning process when used in the classroom. This structure can look differently depending on where you are within the framework. The first level in SAMR is substitution which occurs when you substitute a

2 Puentedura (2013). SAMR: A contextualized introduction. Presentation given October 25, 2013 at Alberta Charter Schools Conference. Retrieved from http://www.hippasus.com/rrpweblog.

classic activity such as reading a biography with watching a video biography or instruct students to use a web tool to research rather than use a print-based encyclopedia. The second level, augmentation, progresses technology integration by involving applications in the learning process, such as highlighting, underlining text, and inserting images or shapes (Chell & Dowling, 2013).[3] Augmentation focuses on technology, not just being used as a substitute but as a functional improvement (Puentedura, 2013). The third level, modification, is achieved when you use technology to involve your students in learning (Hos-McGrane, 2011).[4] The final level, redefinition, is achieved when your students use technology to create something new or complete previously unimaginable tasks (Puentedura, 2013). The following examples are best practices from various educators within each level of this model.

Substitution

As a teacher, you may be hesitant to introduce technology into your classroom but having a way to do it makes it easier for the implementation process. Using SAMR allows you to have a guide to deliver instruction in various ways. A basic example of the substitution phase in practice is instructing students to complete their work on a Google Doc ™ versus using a worksheet. If you are starting this process at the beginning of the school year, the first thing you would need to do is evaluate and assess what technology you have access to in your classroom. Once that is determined, you can decide what types of interactive web applications or

3 Chell, G., & Dowling, S. (2013). Substitution to redefinition: The challenges of using technology. Retrieved from http://shc-edutech.hct.ac.ae/elearning/ed-techseries/articles2013/Chell- Dowling2.pdf.

4 Hos-McGrane, M. (2011). The SAMR Model. Retrieved from http://www.maggiehosmcgrane.com/2010/03/samr-model.html.

websites to use to help with classroom instruction. To reduce potential overwhelm for you and your students, start with one central platform, such as Google Classroom™ , and use up to four interchangeable other supporting applications. These supportive platforms could include interactive applications such as Padlet , Google Docs™ , Google Form™ , and/or Quizizz ™ . Integrating these digital tools into your lesson can help increase the student's academic performance. As you go through the substitution phase, you can provide clarity in intricate lessons that incorporate technology usage, which will lead to the augmentation phase.

Augmentation

The SAMR model's two lowest levels, substitution and augmentation, can work together simultaneously to enhance student learning. As you enter the augmentation phase of SAMR, you begin to use technology to add to the learning process as a functional improvement. This supports students in having a clearer understanding of a complex topic or making lessons more engaging for students than using traditional methods. An example of the augmentation phase in practice would be using a word processing document by inserting a hyperlink to the document, which can lead to another resource or activity.

Modification

As you become more familiar with technology, you can consider moving into the modification level of SAMR by beginning to transform the learning within your classroom. The goal is for the student to see that their work is prevalent not only in the classroom but also on a broader spectrum. For example, having student students create a blog for others to view their work.

Another example of a teacher in the modification level would be using Google Classroom™ or Schoology as an approach to take care of logistical tasks such as messaging students, tracking grades, creating a calendar, creating assignments, etc. Typically, educators on this level will determine that they are all learning new things due to the change in how learning is occurring. In my experience coaching various teachers on the SAMR model, one teacher shared:

> *"I'm learning with my parents and students. I am indeed in a state of duality as I am a teacher and a student. As a student, I'm learning this new way to approach and engage in education. I suppose I'm in the modification part, as I am modifying lessons, modifying how I think about the lessons I prepare and present for a greater audience."*

As change is going to happen, you will come to realize that you are novices together. Students, teachers, the school district, and other stakeholders will, in turn, lead you to want to continue transforming learning.

Redefinition

Redefinition is the level where the impossible can happen in the classroom. This instructional approach is a way of thinking and teaching in which a task or lesson could not be completed without integrating technology. Another teacher I have coached shared:

> *"I set my students up to interact with one another by receiving feedback from their Flipgrid [Microsoft Flip™] posts. Peer tutoring was a daily experience for my students, from them communicating with one another on Google Hangout ™ . I give them time to work together and socialize. When teaching a lesson, I would model it first, and then I would have them work with their partners, complete word problems, and let them know when we would come back together as a class. The students would take over modeling the problems through creating a presentation via Google Slides™ , Padlet™, Jamboard™ , etc."*

When using technology to redefine the classroom's learning experience, it sends a clear message to all students. In the redefinition phase, you connect their learning experiences to their rapidly changing world outside the classroom walls. This instructional practice supports changing an educator's mindset, which essentially changes the student's mindset to transform the learning experience.

Conclusion

The SAMR model can be used when using technology to deliver content and strengthen student relationships. When beginning a lesson, determine three to five advantages of using technology in the classroom. Use these advantages or strategies to complement the lessons you will teach. When preparing for the instructional environment, keep the end goal in mind. The focus of the end goal should be to identify and describe student learning objectives and support the use of technology within your lesson. After completing all of this, it will be time for you to reflect. When reflecting, ensure that you evaluate what occurred in your lesson when technology was used. If needed, revise those strategies to improve the experience and learning outcomes for the students.

5 TIPS TO SUPPORT IMPLEMENTATION

Use the SAMR Model as a Guide

Think of this model as a technology tool that you can use. Find the right tool and use it for the lesson you plan to teach. Determine if technology will enhance or transform the learning process. If it does, then use technology in the lesson. If not, then go for the traditional approach.

Advantages of the Technology

Determine the advantages of using technology in the classroom. Use those advantages to accomplish two things: (1) to identify and describe student learning objectives and (2) to create assessments within the lessons that will be used with the end goal in mind.

Transformation isn't Always Necessary

You don't always have to transform student learning. A few technology additions to an already practical lesson will make a difference in the lesson. Also, know that it is okay to be in the enhancement phase between substitution or augmentation.

Collaborate with Others

Collaborate with your peers within the same and different content areas to broaden your scope of thinking. Creating lessons across various contents allows students to create spaces and platforms to show their thinking.

Opportunities for Transformation

Find opportunities to use technology that exposes students to local and global interactions to showcase their learning. This will help build their cultural understanding of others, how to use technology etiquette, and build literacy skills.

ABOUT DR. AFFTON COLEMAN

Dr. Affton Coleman is a best-selling author, national presenter, elementary math specialist, curriculum writer, academic coach, educational consultant, and academic intervention implementation specialist. She earned her Bachelor's degree from the University of Missouri-Columbia in 2003. In 2015, Dr. Coleman obtained her Master's degree from Prairie View A&M University in Educational Administration. She completed her academic journey in 2021 by earning a Doctorate in Executive Educational Leadership at Houston Baptist University. She began teaching in 2008 and works as an Intervention Implementation Specialist. In her years of education, she mastered her craft and love for teaching using the 5Ps: Proper Preparation Prevents Poor Performance. She dedicated her time to any child she encountered by meeting them where they were using an innovative teaching style.

Dr. Coleman has presented on technology integration at various conferences such as the Educator Rising National Conference, the Teaching and Learning Symposium at Houston Baptist University, TCCA Conference, RTI Regional Conference, HAABSE, TABSE, & NABSE conferences, and other professional development sessions.

References:

Chell, G., & Dowling, S. (2013). Substitution to redefinition: The challenges of using technology. Retrieved from http://shc-edutech.hct.ac.ae/elearning/ed-techseries/articles2013/Chell- Dowling2.pdf.

Hos-McGrane, M. (2011). The SAMR Model. Retrieved from http://www.maggiehosmcgrane.com/2010/03/samr-model.html.

Puentedura (2013). SAMR: A contextualized introduction. Presentation given October 25, 2013 at Alberta Charter Schools Conference. Retrieved from http://www.hippasus.com/rrpweblog.

ERIC LEVELS

"Turn the pain of Learning Management System (LMS) integration into progress."

- Eric Levels

Chapter 3

The PTSDs of the LMS: Parent, Teacher, Student, District Benefits of LMS Integration

by: Eric Levels

For those who know what an LMS is, you may already be shaking your heads and completely turned off by even mentioning the three-letter acronym that may be a four-letter word in your mind. Learning Management Systems (LMS) came into full swing once we were forcefully thrust into a virtual world of sole digital instructional delivery. Education leaders who once were invited to the digital equity table found digital engagement optional and cute, but COVID-19 conditions demanded that education take notice of this being the only option. Now I understand that at first glance, people may think online learning is a game or counterproductive, but there are significant advantages to using LMSs.

Sure, using LMSs may cause a great deal of Post Traumatic Stress for students, faculty, and parents alike. However, I want to share all of the beauties and advantages that can be afforded by highlighting the PTSDs (Parent, Teacher, Student, District) benefits of LMS integration.

Parents

Now, I know that, for the most part, most parents cringe at the thought of students spending extended periods on devices, but imagine that time being engaged in educationally aligned activities. Parents can receive notification of grades and reporting of assignments. Parents/Legal Guardians can know when an assignment is due and if an assignment is also late or has not been turned in. The parent can also communicate with the teacher through the system to discuss further needs and questions regarding assignments or tests/quizzes. The beautiful aspect of the system is that mastery, as determined through standardized concepts, can be tracked over time. For example, if standards are aligned to questions and activities assigned by the teacher, the parent can determine strengths and weaknesses that should be targeted. There is also the feature of determining the number of absences of students if school districts choose to use this feature in the system. However, various school districts may select other systems, so this can potentially be unreliable.

Parents may also find comfort in utilizing the feature of upcoming assignments, which addresses the dreaded answer to the question of what homework or assignments students have. Many students answer, "I do not have any homework or assignments due." Parents can determine what is upcoming if teachers assign due dates to assignments, tests, quizzes, or projects. The calendar view is also available to view and identify the type of assignment and share the full degree of load each day or given week from all subjects. Determining assignment load is only possible if the district or teachers within the school are using and assigning activities within the system with fidelity. The earlier features are also available if you have multiple children and need a clear digital perspective of what is being accomplished in class.

Teachers

Now, teachers, you should have been first with all you have endured regarding online engagement. I know that you have been up to your ears in programs and online tools for educational delivery, but I want to pique your interest or rekindle the usage of a learning management system. I know that you may already be familiar with your district's LMS, whether you want to be or not. I will not inundate you with all its intricacies or all it can do, but I want you to consider a few innovative aspects.

There is an essential need for teachers to automate, when possible, curriculum intervention and choice and voice in learning. I suggest leveraging a tutorial or intervention digital space in your LMS. For example, use the quiz or assessment component to provide immediate feedback to students in the form of multiple choice. Give the students unlimited attempts on the activity, and if they choose an incorrect response, embed a helpful youtube or concept video to review or assist with selecting the proper answer. If the student answers correctly, provide an extension video or concept that enriches or deepens their understanding. The approach, as mentioned above, is excellent for having the students review for major exams and allows for students to review in a self-paced methodology, and is not restricted to the school day. Students also can take the option as many times as needed.

I would recommend setting up an asynchronous tutorial folder in your online class, which is filled with options for students to receive additional assistance from another educator on the concept you have vetted that is aligned. I am sure you have heard people say that they have said something to another person many times, and it was not until the person heard it from someone else that the message was understood. The same can ring true with students who need to engage in comprehension from another educator. You can give them this opportunity by setting up a virtual space of exceptional educators who echo your pedagogical sentiments.

The beautiful aspect of setting up opportunities such as digital intervention, tutoring, and extension is that you can create it once, then use and enhance the material in subsequent years. The LMS allows for enhancements or edits to the lessons whenever you wish to make them. It allows for more flexibility and time-saving options for years to come. In other words, you will not have to start from scratch at the beginning of the year but take what was accomplished and extend it to the next year for the next set of learners. For example, when having students take a test or quiz, depending on their score. Set up a Did not meet, Met, and Mastered lesson area for students and then assign students to the folder that corresponds to their score content comprehension. Setting up intervention and extension opportunities is awesome as other students will not know their engagement is different yet equitable. All students will work on a level consistent with their comprehension of the given content.

Students

The LMS is one of the best means of preparing your students for the present and future of education. Pencil and paper are all but a figment of our imagination. In this digital age, more assignments are being created through project-based learning and digital tools such as Microsoft Word ™ , Powerpoint ™ , Google Docs ™ , Google Slides ™ , and a plethora of online means. These previously mentioned tools can all be utilized through the LMS platform. The platform can also connect to other applications such as Google Drive ™ , which allows you to embed documents or other created activities that have been stored. The opportunities for choice and voice in assignment mastery and delivery are endless through the LMS. Teachers can now provide audio, video, or multimedia delivery options in educational concept understandings. Students of all ages can now turn in assignments, even if they do not know their home-row keys, and teachers can now analyze content mastery at all levels and track learning and growth. Students can also identify their learning deficits and work towards content comprehension through mastery tracking.

Another advantage is that peer partnership can be fostered through the platform. Students can chat with each other about assignments and request clarification or assistance from the instructor within the platform about directions or processes. Students also have the potential to plan their week as well, as there is a calendar view that shows assignments, tests, and quizzes for all subjects if teachers assign dates to those activities within the system. Students can also receive immediate feedback from their peers and teachers in an LMS platform, as students can use rubrics to provide constructive criticism and dialogue through discussions of the material. Teachers can provide grading, give feedback and leave individualized comments that can only be seen by that student. Students can provide additional comments to the thread that only the teacher can see to continue further collaboration and productivity of assignments. Students can also complete multiple attempts on an assignment which allows for the potential for higher scores on failing attempts.

Districts

Districts have the opportunity to leverage LMSs to prepare students for academic advancement and higher education. Online universities and colleges use LMSs to deliver curricula and assess comprehension. Students who use LMSs in K-12 education naturally transition to usage in higher education.

Districts should embrace the use of LMSs because they also can address potential staffing issues. Teachers are essential to the education process, and activities and assignments can be placed in the LMS when there are staffing shortages. A substitute can actively monitor behavior while students can continue content pathways set up by specialists. Even if teachers are to take time off or call in sick, an LMS can allow teachers to set up lessons so students can continue learning.

Master lessons can also be created and copied by other content teachers and enhanced. The idea mentioned above works well for new teachers who need model lessons for a start. LMS modules also assist with proper student assessment across the district to determine what teachers need more support and development by tightly aligning assessments of students to compare apples to apples as it relates to assessment or assignment. LMSs are also leveraged for administration and teachers concerning staff development. Modules and lessons can be created to instruct staff in enriching, enhancing, and extending their educational delivery and practices.

In conclusion, all stakeholders have much to gain regarding the advantages of using an LMS. Disadvantages indeed exist, but it is crucial to maximize the benefits of innovation rather than remain reluctant and reliant on rudimentary methods. The PTSDs of LMSs are perceived as painful but can reveal healing amid engaged difficulty. Engaging and learning something new has never been easy, but the depths of rewards in LMSs, realized by (P)arents, (T)eachers, (S)tudents, and (D)istricts, outweigh the stagnant waters of the familiar.

5 TIPS TO SUPPORT IMPLEMENTATION

Parents: I Can See You

Parents can use their access within the Learning Management System to determine if students are completing assignments and where students are in lesson comprehension.

What Do You Know?

Teachers can now measure content mastery through digital means by formatively assessing student work completion.

Show and Tell

Students have the option of submitting and showcasing lesson comprehension through their own choice and voice in the LMS, and track their own progress.

District Hunger Games

School districts must leverage the full use of the LMS to prepare students for collegiate assignment engagement and maximize limited campus staffing.

Gains over Groans

Places of academic learning should embrace the possible gains experienced through LMS integration rather than the groans of reluctant adopters.

ABOUT ERIC LEVELS

Eric Levels is the Executive Director of Digital Learning for Aldine ISD. He served as the Coordinator of Online Learning, facilitating Learning Management System implementation and training districtwide to over 3,000 staff members and 65,000 students. Levels have worked as creator and curator within the Professional Learning department facilitating training and development of the district through evaluation, education, and elevation of high-yield teaching strategies. His motto is to live life on purpose.

CONNECT WITH ERIC LEVELS

@EricLevels

@EricLevels

@Eric.Levels

EricLevels.com

"Technology is the catalyst that allows you to maximize your instructional time to design inclusive learning experiences to meet the needs of all learners.

– Jocelyn "Dr. Mac" McDonald, Ph.D.

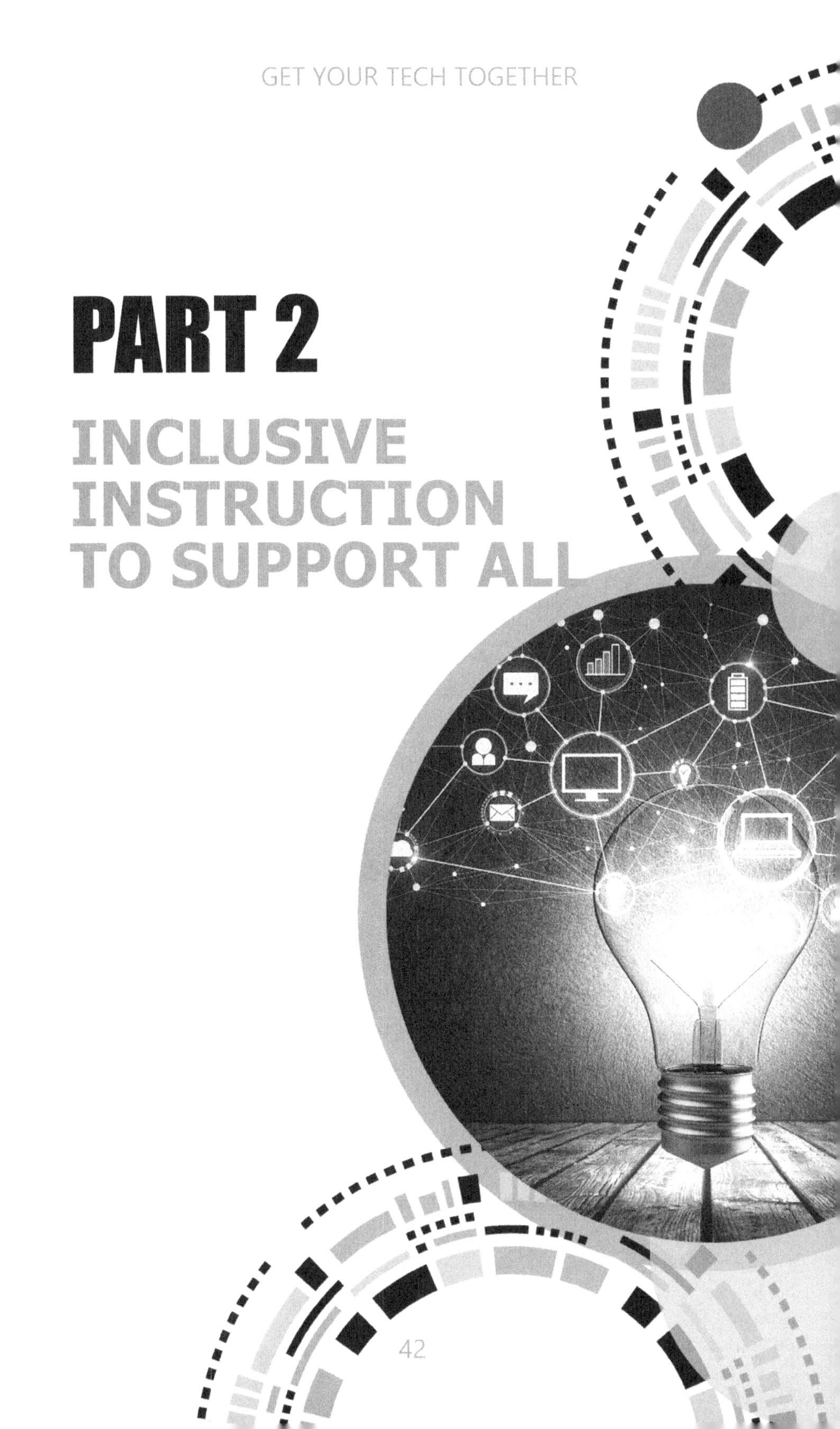

PART 2

INCLUSIVE INSTRUCTION TO SUPPORT ALL

SHERRI POWELL

"As long as we engage students with the modality in which they best learn, all students can learn and are capable of learning."

- Sherri Powell

Chapter 4

Accessible,Differentiated & Personalized Learning for All

by: Sherri Powell

I began my teaching career as a middle school, multi-grade, inclusive teacher in Long Beach, California. It was the first class of its kind in the Long Beach Unified School District, and no one could tell me I wouldn't make a difference! I was a new teacher in a new state, and even the most comprehensive education courses couldn't prepare me for my first real teaching experience. Looking back, I suppose I should have asked what "inclusive" meant and if there was a curriculum to teach the various nationalities, grade levels, students with learning disabilities, and behavioral problems. I found out quickly there was not. I cried nearly every day from frustration, exhaustion, and feeling overwhelmed by all the newness around me: new job, new home, new cultures, and challenges. I didn't quit. I decided to embrace teaching and work hard to become good at it. I had a light bulb moment, and my purpose for teaching became clear. I was on a mission to learn how to engage students regardless of nationality, learning deficiency, background, or behavior.

In this chapter, I will share tips and strategies to personalize and differentiate learning for your students. I will also discuss the importance of making learning accessible for ALL students.

Personalized Learning for All

Consider a class you teach or a classroom you support. Consider critical moments when you see the need to personalize or differentiate student learning. Is it when some students are ready for the next challenge? In this case, they may have mastered the content that supports the learning objectives and demonstrates their understanding of the content. Or what if more than half of the class has some learning gaps? They may show a critical need for more support in achieving the objective. And how do you support the disengaged and those who have mentally checked out?

As educators, we are responsible for meeting every learner where they are. Personalization of high-value learning experiences is not an option. Nevertheless, when planning for personalization, we must consider flexible and accessible content and tools, student agency, and formative assessment. Technology makes it easier for us to work with students on developing and enacting learning pathways, but it also gives students more flexibility in presenting their work.

Imagine providing students immediate access to resources and assessments tailored to their needs. By considering their strengths and weaknesses, students can access what they need to learn, and you can quickly gain insight into how your students learn through various online tools.

To Differentiate or Not? That is the Question

Differentiation requires formative assessments that are effective and timely. For example, teachers use formative assessments to diagnose an individual student's ability to learn instructional strategies that match the individual's readiness level for each course. To support differentiation, combining assessment data with the knowledge of students' interests and needs is essential. For example, Nearpod™ lessons or Google Forms™ interest survey would be a great resource to determine how and when to differentiate for a student.

Supporting Student Agency with Technology

Student agency involves learning through activities and lessons that are purposeful and relevant to all learners. These activities can drive students' interest and are self-initiated with appropriate guidance. As a result, student agency provides some voice and choice in learning.

Often, we overlook student choice when planning our lessons. Nevertheless, motivation is high when students have opportunities to make their own decisions. Don't worry about losing control when allowing student agency in your classroom. It's important to know that student agency does not mean students make all of the decisions. The tools and strategies mentioned below are designed to help teachers engage students and make them look forward to the next class meeting.

Engaging Learning Activities

Learning menus, choice boards, playlists, and HyperDocs provide differentiated and personalized learning opportunities and allow student agency. Other commonalities between these learning activities include:

- They are student-centered.
- They allow students to incorporate their interests.
- They encourage students to interact, collaborate, and create.
- They are favorable for various types of classrooms—including 1:1 technology.

Learning Menus and Choice Boards

Learning menus and choice boards offer students the opportunity for differentiated learning and exploration. Students are provided with various activities in which they can choose to demonstrate an understanding of a topic or unit of study. When students have options, teachers typically find their students more engaged and produce higher-quality work. Many of us may have used some type of choice board in paper format, but with all of the technology tools currently available in the classroom, why not move from paper to digital?

Consider your student's learning needs and access to digital technology. You can create choice boards in many different formats and at different difficulty levels. You can use Google Slides™ to create and organize any of the examples listed below:

- Tic-Tac-Toe
- List of activities
- Bingo Board
- Restaurant-style menu
- Hyperlinked slide presentation

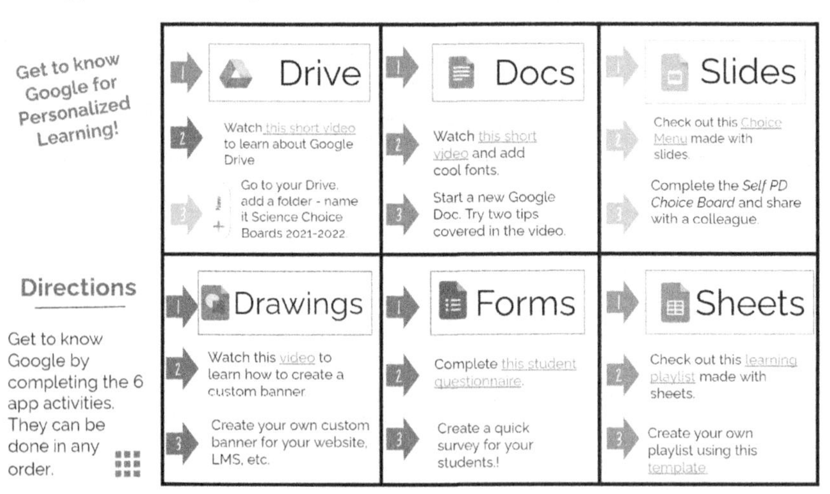

The example below was created using Google Slides ™ to introduce students to Google Workspace ™ applications. Students can complete tasks at their own pace as they navigate through each app.

Playlists: A Path to Personalization

Playlists can be a great path to personalization. They consist of a sequence of resources or activities for students to complete. Playlists help students learn at their own pace by breaking down concepts and units into smaller, manageable tasks. Activities can be arranged to create playlists that reflect each student's academic interests and needs. In many cases, playlists work well in blended-learning environments where teachers instruct individual or small groups of students.

In contrast, others work on playlists that help reinforce concepts learned in class. Below is an example of a 4th grade math playlist that focuses on rounding whole numbers. This spreadsheet provides students access to mandatory tasks and choice tasks. Using the drop-down menus on the spreadsheet, students indicate their status (e.g., Completed, Skipped, Still Working, Need Help).

Enter your activities at the top for students to choose one or more of each color. You can embed this playlist into a Station Rotation or use as a Flipped assignment checklist. Share the spreadsheet with your student(s) with Commenting rights. This will allow them to mark their status (Completed, Skipped, Still Working, Need Help) using the drop-down menu without erasing your content. You can edit the status listing using the data validation feature. They can also comment with a response (i.e. demonstrate their learning).

Directions: You must complete at least activity per color and the **MUST Do's**. **Due Date:** 00/00/0000	**4th Grade - Rounding Whole Numbers:** Learn "Rounding whole numbers to nearest hundred" at Khan Academy.	**4th Grade - Rounding Whole Numbers:** Learn "Rounding whole numbers to the nearest thousand" at Khan Academy	**4th Grade - Rounding Whole Numbers:** Learn "Rounding whole numbers word problems" at Khan Academy	**Must Do**. Rounding Whole Numbers Challenge: Take the challenge!	**FlipGrid** Go to [illegible] Would you rather? Choose an option. Make the choice that gives you the most value. Defend your	**ThatQuiz:** Go to www.thatquiz.org Click on any area to test your math skills.	Enter your activity	Enter your activity	Enter your activity	**Must Do**. This is usually where students demonstrate their learning.
Student 1	Skipped	Skipped	Completed							
Student 2	Completed	Skipped								
Student 3										
Student 4										
Student 5										
Student 6										
Student 7										
Student 8										

HyperDocs

HyperDocs are digital documents in which all learning cycle components can be gathered into one place. Think of it as a one-stop shop! Using HyperDocs, lecturing time can be reduced, and students can access various content in multiple formats, such as audio, video, or text.

A HyperDoc can be explicitly created for a student's individualized needs. What I love about HyperDocs is the ability to personalize student work with differentiated content—without others knowing. Students can access HyperDocs, which may have the same appearance, but are linked to different activities or tasks based on their needs.

Menus, choice boards, playlists, and HyperDocs are all ways of answering these questions:

- What are your students' learning needs?
- How will they engage in the content?
- How will they reflect on their learning?
- How will they demonstrate their learning?

Sometimes the best thing to start with is the easiest. You can use Google Docs™ to create and organize any of these modalities for your students. There are multiple ways to structure them. You are only limited by your imagination. Sure, it's not as flashy as some web-based or paid applications or services, but Google Docs™ is functional and a great way to start for beginning users—teachers and students!

Feedback—It Goes Both Ways

Do you ever provide your students the opportunity to self-reflect on their learning? Just think of the information you could collect and use to better inform your instruction. Having your students engage in collaborative "whiteboarding," interactive assignments, or completing short surveys as exit tickets can help you gather valuable feedback. Some ideas for questions to ask your students might be:

- How would you rate your learning this week? Explain.
- Summarize your favorite lesson this week. Explain why you liked it the most.

- What can I do to improve your online learning?
- Which part(s) of the lesson gave you the most trouble?

For example, these questions would enable you to provide timely feedback that could motivate continued learning. After students have explored, created, and communicated their ideas using HyperDocs, they'll have the opportunity to reflect on their learning, use checklists and rubrics to evaluate their work, and set new learning objectives.

When we ignore real-time feedback, such as adding audio or written comments in a Google Doc™, students and teachers miss essential learning opportunities. Remember that students learn best and are more motivated when feedback is frequent and immediate.

Here is an example of how a 45-minute class period can be divided into sections to suit all student needs:

Time	Virtual	Face to Face
Opening 8:00-8:05	**Whole Group** Attendance/Housekeeping Warmup	**Whole Group** Attendance/Housekeeping Warmup
8:05-8:15	**Whole Group** Lesson Overview Digital Tools • Google Slides • Google Classroom • Pear Deck • Nearpod	**Whole Group** Lesson Overview Digital Tools • Google Slides • Google Classroom • Pear Deck • Nearpod
8:15-8:25	**Small Group w/Virtual Students** • Next Steps • Re-teaching	**Independent Work** • Reading/Writing • Activity • Drill and Practice
8:25-8:35	**Independent Work** • Activity • Vocabulary • Reading/Writing • LMS	**Small Group w/F2F Students** • Anchor Charts • Key Concepts • Next Steps • Re-teaching
8:35-8:40	**Whole Group/Independent Work** • Exit Ticket • Complete Assignments	**Whole Group/Independent Work** • Exit Ticket • Complete Assignments
Closing 8:40-8:45	**Whole Group** Lesson Wrap-up/Review	**Whole Group** Lesson Wrap-up/Review

As you can see, it is possible to implement various digital tools and strategies in the classroom or virtually (e.g., Pear Deck™, Google Classroom™), as well as find time to re-teach certain concepts to engage students in both whole-group and independently.

Make Learning Accessible

There have been significant shifts in K–12 education. We have moved from a traditional, one-size-fits-all model to offering more flexible learning approaches to our students—such as voice and choice—in how they will engage in the content, reflect on their learning, and demonstrate their understanding. Our challenge is how do we make our learning accessible and engaging while meeting the needs of all students?

The very presence of accessibility in a student-centered environment enables personalization. Learning management systems (LMSs) like Schoology and Canvas have built-in accessibility features that allow all learners to incorporate visual representation support like screen readers or speech-to-text tools. For example, not all students who struggle to read or write struggle with the same specific skills. One student may need language support, such as vocabulary development, while another might need text read aloud to them.

Suppose students are given a choice of texts for an independent project, but only one has audio support. In this case, students with difficulty reading (e.g., dyslexia) are limited in their learning. Students who require accessible content can benefit from free, interactive tools such as Microsoft's Immersive Reader or Helperbird for Chrome. These reading comprehensions and learning tools make the text more accessible to students of all ages and abilities.

There are many free and user-friendly accessibility features and tools that you can use in your teaching. Making personalized learning environments accessible to students with learning disabilities is key to ensuring they have a greater opportunity to exercise agency. Ask yourself, "Is 'x' available for every student?" and "Can every student use 'x'?" If your answer is "No," it is essential to make your lessons more accessible for your students. It should not be an option.

Invest Time and Presence in Professional Development

Substantial professional development is essential to differentiate and personalize instruction for students. Like our students, not every teacher learns the same way, and having the tools and support is crucial for successful implementation.

Equipping yourself with the necessary tools and skills for teaching and learning could help improve performance.

5 TIPS TO SUPPORT IMPLEMENTATION

Design Learning to Engage Every Student

Find out what your students' interests are. Students tend to engage fully when their interests and needs are met.

Plan for Accessibility and Differentiation

When each student has their own copy of a playlist, HyperDoc, or learning menu, customize each copy to meet their needs.

Join a Professional Community

Leverage professional learning communities (PLCs) as a great source of on-time learning and collaborations with your peers. For instance, joining professional learning networks such as edWeb.net makes it easy for you to collaborate, engage in peer-to-peer learning, and share ideas and resources.

Make meaningful connections

Take advantage of the small opportunities to reach out and connect with individual students. Engaging inside conversations to check in and see how they are doing is okay. This helps build trust and communication skills between you and your students.

Be a forever learner

Engage with materials. Collaborate with your It's okay if your peers, students, and parents know that you don't know everything. We learn best from each other and should not be afraid to show our vulnerability and weaknesses. Be open to new ideas and strategies. Collaborate with your peers. Find opportunities for extended workshops and specialized training sessions that lend towards accessibility, personalization, and differentiation strategies.

ABOUT SHERRI POWELL

Sherri Powell's 23 years in education, passion to empower, dedication to innovation, and show-no-fear approach to technology makes her a positive change agent for education. She teaches and transforms technologically timid teachers and administrators into fearless educators who embrace the latest tools and resources to teach and engage their students. In addition to holding an Ed.S in Technology Management and Administration and an M.A. in Technology Education, Sherri is a Google Certified Trainer and Microsoft Innovative Expert and Trainer. At Region 4 Education Service Center in Houston, Texas, Sherri designs and delivers engaging professional development experiences. Additionally, to integrate diverse technologies with her innovative teaching methods, she is constantly seeking new ways to reconfigure the traditional classroom into a place that fosters active learning.

DR. ELENA SILVA-LEAL

"Knowledge emerges only through invention and re-invention, through the restless, impatient, continuing, hopeful inquiry human beings pursue in the world, with the world, and with each other."

- Dr. Elena Silva-Leal

Chapter 5

Shifting the Paradigm of Your Instruction

by: Dr. Elena Silva-Leal

Thursday, March 12, 2020, changed the lives of students, teachers, parents, and administrators in my district. After school, we received text messages, phone calls, and emails informing us that due to the rise in Covid-19 cases in our city, we were not to report to school until further notice. The next day we were allowed to collect laptops, monitors, and any teaching materials we needed. The following week campus administrators met over Microsoft Teams™ . They drafted a campus plan for online instruction, including distributing hotspots to students, protocols for grading and attendance, and a crash course for online teaching. The remainder of 2019-2020 was one-way virtual. The following year, teachers provided instruction virtually to students and face to face in the classroom. Teachers prevailed despite the struggle to juggle the two audiences and teaching modalities. That success was due to establishing routines and protocols that made learning accessible to all learners and planning the integration of technology tools with purpose.

I started with this introduction because this launched a paradigm shift for many educators. Technology forever changed the planning and delivery of lessons and how students respond. In this chapter, I discuss instructional practices to support the paradigm shift of integrating technology into your instruction.

Managing Equity in Student Voice

As educators, we think about and plan authentic learning experiences that require students to think deeply about how they interact with their world. As an administrator that serves teachers and underserved students in an urban school district, we want our students to bring all of their experiences and connect what they learn beyond the walls of the building to what they learn in the building. At our campus, one area of focus is contributing voices; we want everyone to speak. We have to get creative in how we encourage and manage discourse, and technology lends itself to support these efforts. One tool that we introduced and continue to use is the *Wheel of Names*. It is a web-based application we used for "cold-calling" to ensure equity in voice in the classroom. You can simply sign in using your Google account or email address. Then, enter or paste student names into the 'entries' box, and the wheel is ready to use. We use the wheel to ensure equity in the voice of students as it randomizes opportunities for anyone and everyone to speak. But wait, there's more. Some of the teachers opted to have students submit writing topics, then the teachers added them to the wheel. The students loved seeing their ideas on the wheel, adding a little more excitement to the writing process. The math teachers used the wheel with four options, (a), (b), (c), and (d). On the board were four equations; where the spinner would land is the equation they solved, and that is how they built excitement. Now, of course, there are various technology tools that you can use to support the same effort. In its use as a randomizer option, the technology aids in ensuring every student is engaged in the anticipation that they can be selected to contribute to the class discussion. Replacing a running roster or popsicle stick as a tool with a visual such as the *Wheel of Names* builds excitement and energy as students watch their names spin on the screen. Furthermore, the randomizer can shuffle tasks, roles, and assignment-super fun.

Instructional Routine to Support Student Voice, Literacy, and Technology

On the theme of voice, at our campus, it was essential for us to build students' confidence to speak. Whether the topic is about themselves, an issue in their community, a pop culture topic, a memory, or a movie or book review. One strategy that can be implemented to support students in obtaining this goal is an instructional routine for students to use academic language in scripting their monologue that they will digitally record on Microsoft Flip™ (formerly Flipgrid). This routine and tool combination notably supports teaching all of our students to write with authority and speak with command while intentionally supporting our emergent bilingual students with their command of the spoken English language. Students are presented with a probing question, sometimes very specific, sometimes generic. Students, next, draft their assertions and use primary or secondary sources to identify three relevant pieces of evidence that support their argument. This is followed by an explanation of why that evidence supports their claim. Finally, students write a summative statement of their stance. This routine provides scaffolds; paragraph frames, sentence stems, and even academic and the option to use content-specific word banks. After the writing process, we gave students feedback on their responses. Once satisfied with students' writing, we would direct students to read and record them using Microsoft Flip™ . Microsoft Flip™ is an interactive tool allowing video discussion, collaboration, and sharing. Teachers post a prompt then students access Microsoft Flip™ from their phones or laptops to respond to the teacher's prompt and comment on classmates' responses.

To set up Microsoft Flip™ :

1. Create a free account with either Microsoft or Google, then *create a group for each class and name them accordingly.*
2. *Create a topic (or prompt) where students will place their video responses. For example, the structured writing I described.*
3. *Give the assignment a title and add instructions to the description box.*
4. *Copy the generated link and add it to your preferred classroom learning management system (LMS).*

Students can then click on the link and add a response. Use a rubric to provide feedback to students, share a few in class, and, best of all, generate pride in their presence.

These two straightforward digital tools demonstrate just a glimpse of how we thought of and planned for the integration of technology. My goal in supporting students, teachers, and my team has always been to design a learning environment that makes students want to question how things work and wonder about everything around them. I want students to create products that are meaningful to them. In doing so, careful and strategic planning must be considered. A teacher's presence remains one of the most important components of classroom learning as we can emotionally engage our students. When creating a technology-rich learning environment, we engage learners cognitively and behaviorally. This can be done by establishing equity in voice and invoking creativity in thought. Using technology to design, develop, and implement opportunities for students to personalize how they respond to a task allows for student autonomy to respond at their own pace and to the depth of their ability. As students become more comfortable with the software, they will become more creative with integrating technology on their own and develop soft skills that will prepare them for the workplace.

Being Intentional with the Use of Technology

Whether instruction is online or face to face, we are sure

of one thing: we must engage the learner. Emotional engagement describes the level of enjoyment and interest the learner derives from the learning task or skill itself and from being an integral member of the class (Bobis et al., 2015) [1]. When planning, determine how the tool will be used and by whom. Consider the specific goal to be accomplished: is it a tool for the teacher, a tool for student collaboration, a place for student creativity, or feedback and input.

Once you have determined the purpose of using a technology tool, practice implementation. Collaborate with other teachers and explore application features that will support your instruction. Determine the ease or difficulty students will experience as they learn to use the tool and identify barriers and roadblocks students may encounter. You'll be surprised how students are quick to learn and discover elements of digital tools, but they may also cease to persist when the task becomes cumbersome; consider a contingency plan for support. This is also an excellent opportunity to discuss if the digital tool will meet the varied needs of different learners. Consider how students can personalize their products, self-regulate their learning, or collaborate with their creativity.

Planning and practice will happen almost simultaneously. Plan one of two things, whether the digital tool is something you will use during direct instruction or if this is a tool you want students to choose from to complete their student product. Then, begin planning with the standards from your content-specific state curriculum, identify the skill mastery students must demonstrate, and determine what digital tool will provide the means to do so. Collaboration, planning, and practice provide necessary feedback and discussion from the team. Focus on the desired outcome.

1 Bobis, J., Way, J., Anderson, J., & Martin, A. J. (2015). Challenging teacher beliefs about student engagement in mathematics. Journal of Mathematics Teacher Education, 19(1), 33–55. https://doi.org/10.1007/s10857-015-9300-4

5 TIPS TO SUPPORT IMPLEMENTATION

Be Purposeful When Selecting Tech Tools

Consider the specific goal to be accomplished with the tool: collaboration, creativity, or connecting beyond the classroom. Keep in mind that the focus will be learner-centered instruction and experience.

Practice Using the Tool

Be familiar with each technology tool so that integration is polished and composed. Work out logistics and plan catches to clarify potential misconceptions.

Plan, Plan, Plan

Be specific in how the technology tool will support students in mastering the learning objective. Design the learning experience to challenge student thought and to generate student engagement and effort to invest and persist in learning tasks.

Introduce the Digital Tool

Establish norms and a community of practice to successfully integrate educational technology. Do not assume students know how to use the tool; take the time to introduce the technology and for students to explore its use. Always introduce one technology tool at a time.

Repetition

Practice the implementation of the technology tool. Eliminate undesired missteps and improve application over a few weeks. Once the desired level of competency is reached, introduce a new tool and application.

ABOUT DR. ELENA SILVA-LEAL

Dr. Elena Silva-Leal is in her twentieth year as an educator. Her background includes roles as a classroom teacher, department chair, teacher development specialist, and school administrator in private and public schools in urban areas. In her latest role, Dr. Silva-Leal served as a high school administrator in the largest school district in Texas. Her professional development and support of teachers resulted in change and improvement of the learning environment and the emergent bilingual student organizational program. She specializes in support of turn-around campuses and has reached hundreds of teachers from across the state of Texas on inquiry-based instruction, scaffolds, digital integration, literacy routines, and data-driven instruction. She believes students should experience authentic learning opportunities to think critically about their world.

ARELYS BARRETO-PAZ

"We are all English learners no matter our socio-economic background or ethnicity."

- Arelys Barreto-Paz

Chapter 6

Leveraging Technology to Support English Learners

by: Arrelys Barreto-Paz

Technology integration in the classroom has shifted since Covid-19 forced us into a virtual experience. Before Covid, many teachers would play videos from different sources, maybe even use Kahoot!™,[1] or project a Microsoft PowerPoint™ (PPT) during instruction. Some may have added a technology station in the classroom using applications (apps) with ready-made pathways for students. But how effective was that? Were students engaged? Were they all participating? What about special population groups and emerging bilingual students still developing in the English language proficiencies?

Going virtual forced a lot of teachers to use technology in different ways. It ensured students were not hiding behind a camera by expecting students' responses to show on the app. In our district, and I am sure in many other school districts, teachers had to rely on digital resources such as a Learning Management System (LMS), Microsoft OneNote™ ClassNotebook, Google Workspace™, and Pear Deck™.

1 Kahoot! and the K! logo are trademarks of Kahoot! AS.

Using digital tools like these can routinely impact all students, including students who are English learners or Emerging Bilinguals.

We may now see that some teachers are returning to old habits with just paper and pencil instruction, while some may have continued using some of the digital resources but may not be using them to their full potential. Or maybe did not keep up with the app's updates or did not use the app or digital tool properly. Let's take a closer look at different digital tools that make learning engaging and impactful and can be suited for all various learning needs, with particular consideration for students learning the English language.

Learning Management Systems (LMS)

Teachers and students may use a specific Learning Management System (LMS) depending on the school district. It is a place where all assignments, assessments, resources, and curricula can be housed for all. An LMS like Canvas, Itslearning, and Schoology may have standard features like a rich-text editor and an Immersive Reader. Teachers may also use Google Classroom ™ if their school or district does not have a specific LMS.

Rich Text Editors

A rich text editor is typically an embedded feature in an LMS and can be found in other digital applications that support text customization. It has various features that can benefit students in many ways. For example, it provides the capability to insert images, audio, and video to support directions and enhance instruction. Additionally, it can provide ways for students to respond to assignments and discussions and better engage in the learning experience.

Images are great nonlinguistic representations that can

support instruction in various ways, from language support to connecting new learning. When teachers add images to their directions, it helps to see a clear picture of what the expected assignment is about. Using the rich text editor to add visualization can also anchor students' previous learning. When giving writing assignments, it is best to include an image that connects to the topic. It helps the student visualize what they should be writing about and get further information that is missed from the writing prompt. Writing is the last proficiency level developed by many English learners (may also be called Emergent Bilinguals), so the more support and techniques are provided to students, the faster they advance on their level.

Audio is another feature that may be found in a rich text editor that teachers can use as recorded audio of the writing directions or by students to respond to the assignment. Students' listening and speaking proficiency levels are faster to build since it is the proficiency level they get a chance to practice in the classroom more often. It is best to include sentence stems when students are expected to respond using audio as it serves as a template to begin their response. It is also a good practice to use this tool as often as possible and throughout the school year, so students practice recording their responses for when they are tested for the new proficiency level.

Immersive Reader

Using this tool can be beneficial and disadvantageous to students as it depends on the extent of use. Instead of using it as leverage for learning the language, students may misuse it to cut corners while learning the language. Depending on their proficiency level, students should rely less on the tool. If the student is on a beginner level, it is common to see students utilize it as a "reader" of the entire text while translating all to most words. The immersive reader can support students with reading through text-to-speech. This feature can help them listen to how the word is pronounced, broken into syllables, see what part of speech it is, or have a visual representation of the word. It can also support the students who have Dyslexia that needs a tool to help them read the text. We can find this tool in all Microsoft programs like OneNote, Word, Teams, and even other applications like Book Creator™, Microsoft Flip™ (Formerly Flipgrid), NearPod™, or Pear Deck™.

OneNote Class Notebook

Microsoft OneNote™ Class Notebook can be used as a digital notebook for students in which teachers have multiple tools to make learning engaging. Teachers can distribute handouts and structured content to students as assignments and even accommodations for specific students. Teachers can assign group activities and push out extra resources when students do not understand the concept. When pushing out additional resources to students who are emerging English learners, it is essential to keep Sheltered Instruction Strategies (also known as content-based instruction) in mind depending on the English level. For instance, a beginning student would need sentence frames to be able to write about the content with the addition of visuals. As the student advances to a higher level, less support should be given. Instead of sentence frames, an advanced student would get a sentence starter if needed. OneNote has an immersive reader tool, which helps with students' learning. It has a recording audio feature that is a great tool to build student speaking practice or as a way to build accommodations for a student. When providing the opportunity for students to record, keep in mind that beginning-level emerging English learners may also need sentence starters. It can also embed PowerPoint presentations for students to take notes on the side of the slides or add supporting visuals. OneNote is an excellent way for teachers to see students' journals to understand their level of learning and thinking.

Jamboard

Google Jamboard™ is a Google Workspace™ application, a version of a digital whiteboard that can be used collaboratively. It can be shared between the whole class digitally. Any technology application can be challenging to use and implement if there are no expectations or guidance before allowing students to use it. However, teachers can use Jamboard for thinking maps as part of the lesson for students to work collaboratively, with partners, or individually within the application. When students work collaboratively, an excellent way to see each student's work in real-time is by assigning a color to each student, that way, you can see a colorful assignment made by each student. You may have seen this activity in the classroom as a collaborative poster activity on anchor chart paper. This activity can be done in Jamboard in multiple ways using the pen/pencil feature or sticky notes. Jamboard allows students to show understanding, similar to the traditional use of pencil and paper, by incorporating a tech tool that helps them represent their thoughts. You are providing a way for students to create a visual representation of mastery of the lesson by lowering the affective filter in students to exemplify their level of understanding.

Pear Deck

In our school district, Pear Deck™ was one of the many tools used by teachers in our school district used when teaching remotely, with the capability of asking six types of questions to interact with students. A teacher can ask questions where students can enter a written response, choice selection, drawing, numbered, draggable, or a website for students to visit. All of this can be done using the Pear Deck™ application to embed other features and third-party applications so students don't get lost during instruction and are on track with the pace of the lesson. It is best practice to provide an opportunity for students to interact with the lesson with one of the six types of questions every 5-7 minutes of teacher talk. You can easily use Pear Deck™ as a tool to support this best practice with built-in questions and interactivities. Another feature in Pear Deck™ that can help students is teacher feedback during a teacher-paced lesson. When trying to lower the affective filter in students, trying to minimize the anxiety levels by having students respond is a safe space to a topic the student doesn't quite understand. When lowering the affective filter, it is best to use the teacher feedback feature. Students will feel comfortable responding to a question because the teacher will provide feedback in an instant private way. Another way to use Pear Deck™ is at a station in the form of a student-paced lesson when extra time is needed to be given to complete that day's lesson. It also allows teachers to make the lesson interactive with high student responses from all students rather than hearing from the same students all the time.

5 TIPS TO SUPPORT IMPLEMENTATION

Rich Text Editor

Rich Text Editors are available on many learning platforms. It helps teachers provide enough information about the assignment and a way for students to respond depending on their level of mastery or accommodations needed.

Immersive Reader

Use the immersive reader to assist students with reading accommodations such as text-to-speech, visual formatting, nonlinguistic reading support via the picture dictionary, and language translations for dozens of languages. This application will support students with pronouncing words correctly, decoding text, and translating when needed.

Microsoft OneNote™ Class Notebook

Use OneNote as a digital journal where all students receive the same information or modified based on their needs, even when a student misses a day of instruction.

Google Jamboard™

Use Jamboard to help students represent their learning, work collaboratively with others, or show their understanding. It is easily accessible through Google Workspace ™.

Pear Deck™

Pear Deck is a commonly used slide extension where students are kept with the class's pace, and teachers can see how they respond to questions during the lesson.

Bonus Tip

Always include sentence stems and visuals for students to aid with increasing responses.

ABOUT ARELYS BARRETO-PAZ

Arelys Barreto-Paz has served as a bilingual educator for several years and has always sought to integrate technology into instruction to support all learners across all content. She emphasized supporting other colleagues through modeling lessons, coaching, and finding resources to support instruction. She went to school to further support teachers and her students to get a master's degree in instructional technology leadership. She has experience with working with elementary grade levels but also worked with the Multilingual Department to extend herself in supporting teachers and students in shelter instruction strategies. Later, she joined the Academic Instructional Technology Department to follow her passion for integrating technology into instruction while supporting teachers and students. Her experience in instructional technology extends from coaching teachers to managing a team of instructional technology coaches in a 1:1 technology middle school expansion program in a large urban school district serving over 200,000 students. This has allowed all middle school students to have access to technology while receiving instruction that is engaging and impactful.

> "There is a science to teaching that requires strategy, protocols, and frameworks bundled into a network of pedagogy. Let the technology illuminate your practice to enhance student performance.
>
> *– Jocelyn "Dr. Mac" McDonald, Ph.D.*

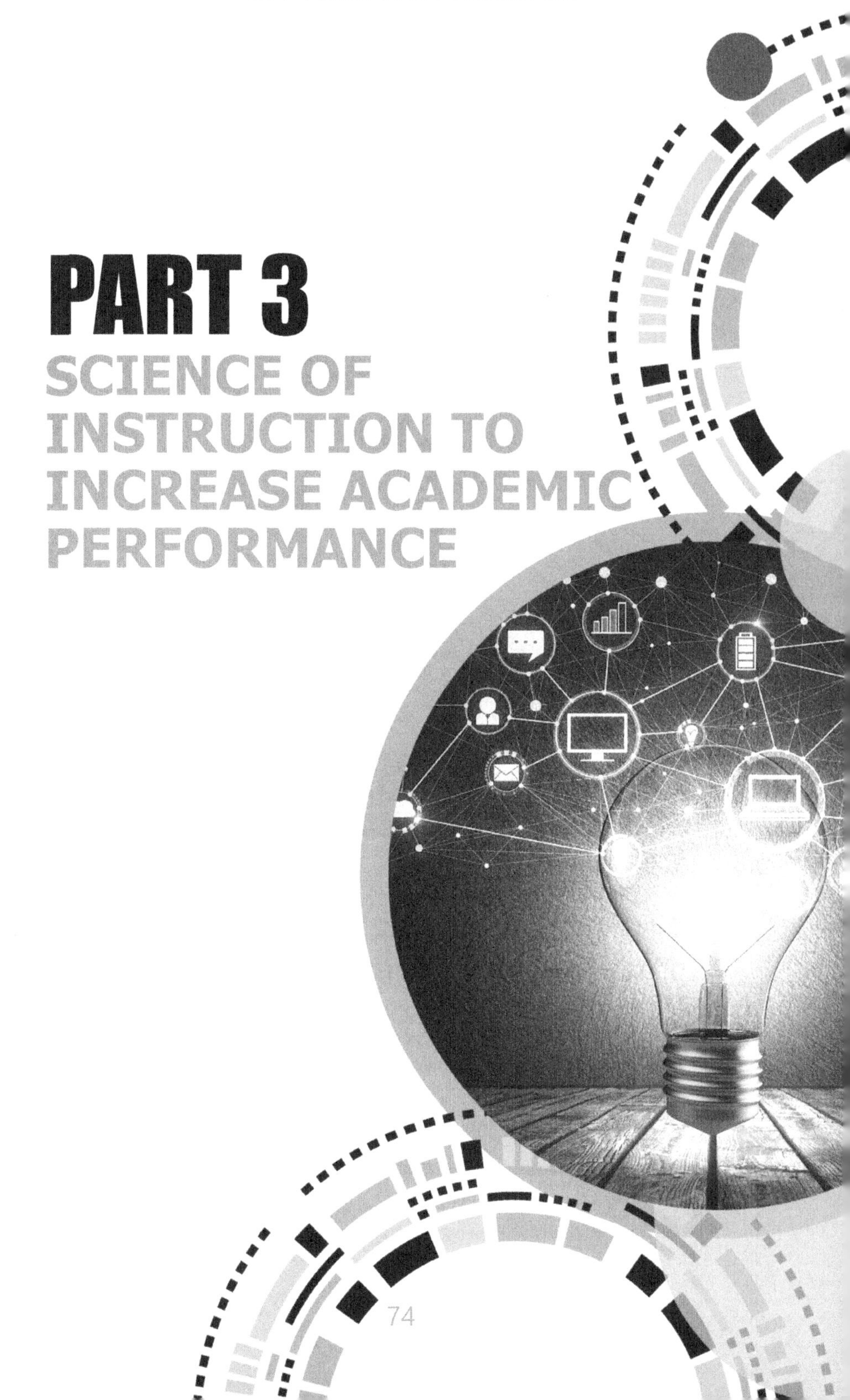

PART 3

SCIENCE OF INSTRUCTION TO INCREASE ACADEMIC PERFORMANCE

CICELY KELLY WARD

"Teaching and learning, like the wash cycle, is a process with no shortcuts. If you don't want half-washed student outcomes, then you can't have a half-washed plan."

- Cicely Kelly Ward

Chapter 7

Data-Integrated Teaching & Learning Model

Using Data-Driven Instruction, Planning, and PLCs for Effective Instruction with Technology Integration

by: Cicely Kelly Ward

Teaching and learning is a comprehensive approach that informs students' performance and skill mastery. If we take all the right actions needed to improve student learning, then why aren't we getting results? That is the million-dollar question! Often teachers and leaders forget that teaching and learning is a cycle and fail to maintain its systematic order. Skipping parts of the process will not help you or the students reach the goals faster.

Let's think about the wash cycle of a washing machine. Initially, you make some baseline decisions pending what is being washed. This is also where sorting and preparation take place. Some laundry may be more soiled than others, requiring additional attention during the wash. Next, you determine the settings. You might ask, "What temperature should the water be, and should I add the extra rinse?" This preparation and decision-making are critical if clean clothes are the desired results.

The wash cycle is an analogy for the Data-Integrated Teaching and Learning Model (DITL Model) (see diagram). The DITL Model consists of Data-driven Instruction (DDI), Instructional Planning, and Professional Learning Communities (PLCs). These are all courses of action that directly impact student achievement. In other words, if teaching and learning had a wash cycle, these would be the significant dials on the control panel. Sometimes instructional adjustments must be made before and after a lesson that requires additional resources for student individualization. However, we cannot answer questions regarding what should be added to instruction until a thorough data analysis has occurred. A detailed analysis will support decision-making around the reteach approach and the appropriate instruction type.

The DITL Model proves to be just what teachers and leaders need to see positive results in student achievement. DDI is a process that connects to curriculum, instruction, and assessment. It is the foundation for a series of steps around creating a data-driven culture of assessment, analysis, and action that should take place during Professional Learning Communities (PLCs). Within the PLCs, consistent team practices must be implemented, including instructional planning, practicing the instructional plan, learning new skills, and data analysis with action planning. The DITL Model can be the framework for establishing school-wide systems. An application to develop school-wide systems is Microsoft OneNote™ . OneNote™ is an online collaborative space that will help compile your data, manage instructional planning, and execute various PLCs for organization and focus. This digital notebook can be an opportunity to collect data about your practices and agreed-upon processes as a school. Leaders and teachers will be responsible for adding resources or expected deliverables to the digital notebook to enhance effective teaching and learning.

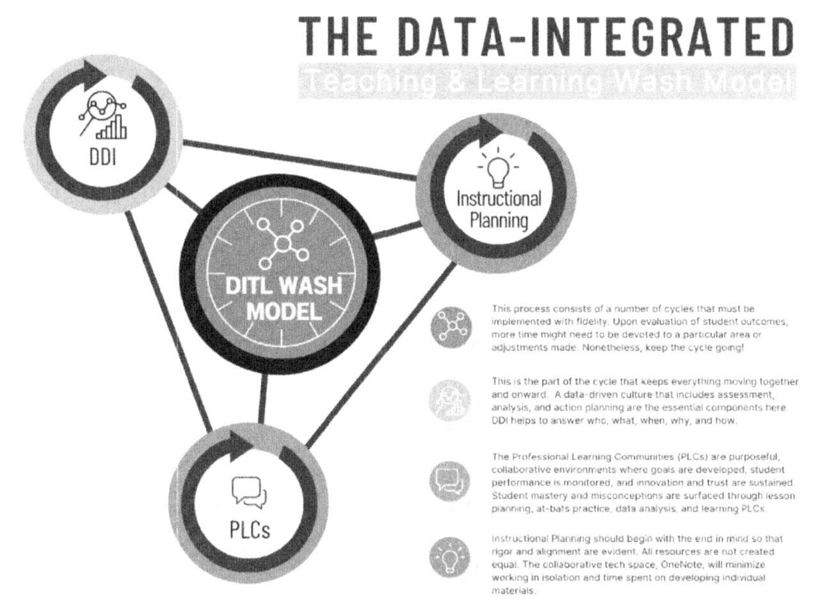

Data-Integrated Teaching and Learning Wash Model C 2022 Cicely K Ward.

DATA-DRIVEN INSTRUCTION (DDI)

DDI is what drives teaching and learning. It serves as a gauge of what instructional support is needed. It also serves as a means to monitor student mastery and misconceptions. I refer to these as Student M&Ms. Just like the candy, m&m's™, attaining student mastery is always sweet, but getting to the core of student misconceptions requires getting past the hard shell. After students are assessed, data disaggregation and analysis should take place. You want to be sure the assessment is accessible so that questions can be analyzed in two ways: by standard and the distribution of student responses (Bambrick-Santoyo, 2019). [1] A fancy platform or learning management system is not needed for data analysis. If you have a report or spreadsheet of the data, then it can

1 Bambrick-Santoyo, P. (2019). Driven by data 2.0: A practical guide to improve instruction. Jossey-Bass.

be utilized in Google Sheets™. A feature in Google Sheets™ called Explore helps you quickly visualize your data (bit.ly/analyzedatainsheets). You will be able to find patterns and trends across the data set. Using conditional formatting to color-code based on a set of criteria is another way to reveal relationships and information in the data. Another benefit of the Explore feature is the capability to represent the data in a graph and pivot table. Using these tools will lead to a more practical interpretation and management of your data.

A deep analysis of the questions and responses is critical to understanding the Student M&Ms. Finally, an action plan is created to emphasize priority standards and address misconceptions with a new teaching approach. Priority standards are determined based on the frequency in the assessment blueprint or whether students have historically struggled with the standard. You also want to prioritize standards to ensure skill-building of before and after grade-level content. Vertical alignment of standards can be a game-changer in a school-wide plan. Data analysis and action plans are useless without timely implementation and follow-up (Bambrick- Santoyo, 2019). [2] Students must be active participants in the process by setting goals, tracking their progress, and reflecting on their learning. This level of accountability helps students take ownership of their learning.

INSTRUCTIONAL PLANNING

Lesson planning serves as the daily instructional agenda to help you organize teaching and learning. Lesson plans should show the main lesson elements from start to finish and be done in a cohesive, structured manner. Many teachers may use Google Docs™ or OneNote™ to plan and structure their

2 Bambrick-Santoyo, P. (2019). Driven by data 2.0: A practical guide to improve instruction. Jossey-Bass.

lessons. There are various examples of lesson plans available, and there is no right or wrong way to construct them. However, consider the following fundamental elements: "hook/ engage" activity, guided practice, and independent practice. The "hook/engage" activity is meant to spark interest and curiosity. Integrating technology into the lesson plan can be a quick and easy way to generate student discourse and set the tone for teaching and learning. For example, video clips and illustrations can stimulate creative thinking and novelty. Once you have the student's attention, the "hook/ engage" can help make connections in other parts of the lesson cycle, which will serve as an anchor for learning. The guided practice provides opportunities for students to make meaning to the new learning. Students need time to grapple with the information so that new connections are made, and teachers need to use this time to model thinking, ask probing questions, and check for understanding. Two excellent online tools to facilitate guided practice are Nearpod® and Pear Deck™.

Both platforms increase student engagement and allow teachers to quickly check for misconceptions in real time. The independent practice will enable students to apply their new learning in various contexts. Again, Nearpod ® and Pear Deck™ are high-quality technology options during this part of the lesson. Polls, quizzes, and open-ended questions can be included to allow students to show their learning through collaborative, interactive experiences. If instant feedback is what you are looking for, then Quizziz™ and Kahoot!™ [3] are ideal options. These two platforms are used not only as game-based learning but also as self-paced learning for students. The power of these tools is their ability to assess student understanding before, during, and after instruction so that

3 Kahoot! and the K! logo are trademarks of Kahoot! AS.

adjustments can be made immediately and intentionally.

Teachers must have time to plan a high-quality lesson the first time it is taught. This means understanding the standards and knowing how they will be assessed. You can determine this by identifying what students need to know and do to show mastery. The best approach for effective lesson planning is beginning with the end in mind. Consider the verbs in the standards and ensure that students are doing that. For example, suppose the standard asks students to describe and illustrate the continuous movement of water above and on the surface of Earth through the water cycle. In that case, teachers must plan for students to describe and illustrate that process. It is also advantageous to review past state assessment items. Teachers can use the assessment items to identify the alignment in how the standard was tested versus how it was written and determine the skills students need to answer the items correctly. Contrary to what many think, this is not "teaching to the test." It guides learning and crafting lesson plans focusing on skill-building, leading to mastery of the standard.

PROFESSIONAL LEARNING COMMUNITIES (PLCs)

PLCs promote teaching and learning and are critical for developing a positive culture of trust, transparency, and commitment among the team. They are designed to have a direct impact on student and adult learning (DuFour & Fullan, 2013).[4] Therefore, the most effective types of PLCs include data analysis and action planning, instructional planning, learning, and practice (Bambrick-Santoyo, 2018).[5] The learning and practice PLCs are the most underused. Learning

4 DuFour, R., & Fullan, M. (2013). Cultures built to last: Systemic PLCs at work. Solution Tree Press.

5 Bambrick-Santoyo, P. (2018). Leverage leadership 2.0: A practical guide to building exceptional schools. Jossey-Bass.

PLCs are intended for teachers to acquire new knowledge or skills. The practice PLCs aim to rehearse a part of the lesson or apply new learning. This is referred to as "at-bats." Teachers can use a collaborative document such as Google Sheets™ or Google Docs™ to provide feedback to their peers during at-bats. The feedback captured will be available for later reference. It can also serve as data collection for the receiving teacher to assist with identifying areas of strength and improvement while delivering instruction. When teachers are given opportunities to practice implementing a strategy or part of the lesson cycle and receive feedback, it supports their professional growth. It also builds confidence leading toward greater self-efficacy.

— "If we want to grow strong students, we must nurture and grow strong teachers."

PLCs are a surefire way to improve student achievement
PLCs are a surefire way to improve student achievement
because they optimize the learning experiences for teachers and staff. Trust the process. PLCs are also cyclical, but teacher collaboration and development are impeded without consistency and authentic engagement. Too often, administrative tasks or school checklists get in the way. If OneNote™ is consistently used, your PLCs will remain active with collective accountability (bit.ly/onenotenotebook). For example, a data page or section would be appropriate for data PLCs. This page could have a shared action planning template that is accessible by everyone to encourage teamwork and participation during the PLC. OneNote™ can reinforce the attributes mentioned above of building a positive culture. Those tasks and checklists are important too but do not belong in a PLC, nor should they be associated with a PLC. The work within PLCs is different, and teachers must know and feel the difference. Otherwise, their perception of

and commitment to PLCs will be damaged, making it harder to build and sustain the PLC culture desired.

The DITL model is a cornerstone that lays a framework for effective teaching and learning. My experiences in the field demonstrate that you can't do one part of the model without the other if the goal is to increase student performance. Like the wash cycle, the key is to figure out your school's capacity to prevent unbalanced loads. You want to be sure to read and analyze the care labels for informed decision-making, especially when it involves your delicates: your students.

As you evaluate progress, front loading, top loading, and occasional pausing will be essential. Nonetheless, trust the process and adjust the settings as needed to get the best-washed results.

5 TIPS TO SUPPORT IMPLEMENTATION

Use the DITL Model

Trust the process and SWEAR by it. Secure the plan. Work the plan. Evaluate the plan. Adjust the plan. Repeat.

Monitor and Respond

If you don't monitor and respond to Student M&Ms (mastery & misconceptions), students will have difficulty reaching the expected target.

Collaborate using Microsoft OneNote™

OneNote™ is an online collaborative space that will help compile your data, manage instructional planning, and execute various PLCs for organization and focus.

Leverage Technology to Access Understanding

The power of implementing technology tools is assessing student understanding before, during, and after instruction so that adjustments can be made immediately and intentionally.

Use Technology to Streamline and Automate

Use technology to streamline and/or automate routine procedures and practices.

ABOUT CICELY KELLY WARD

Cicely Kelly Ward has worked in public education over the past 22 years. She has served in various roles, including teacher, instructional and teacher development coach, assistant principal, data-driven instruction specialist, and manager These roles have afforded Cicely opportunities to coach and build data-driven instruction systems for schools and district staff to inform instructional practices. Being part of school reform and turnaround efforts for the past twelve years in a large urban school district has ignited Cicely's passion for individual and team development. She is committed to providing high-quality professional learning for campus leadership teams, teachers, and district staff to increase student achievement and data literacy, improving school-wide systems and classroom practice, and transforming leader and teacher efficacy. Cicely received her undergraduate degree from Alcorn State University and her master's degree from Sam Houston State University. She is a member of Delta Sigma Theta Sorority, Incorporated. In her leisure time, Cicely enjoys spending time with family and friends, and traveling with her husband.

ERIN CARROLL

"Every child has the right to learn in a way that is engaging and connected to a world they recognize. We must set the bar high for students and give them tools and visual supports to encourage leadership and accountability that will support them in exceeding expectations every time."

- Erin Carroll

Chapter 8

The 5 E's for Effective Literacy Instruction

by: Erin Carroll

Literacy is often disliked by young learners, not because it isn't important, meaningful, or interesting, but because we as educators do not teach literacy in a way that is relatable, engaging, challenging and fun to students. Students can not retain what they don't like or understand, they must connect to the learning. This chapter details the 5 E's of effective literacy instruction: Engage, Educate, Empower, Embed, and Elevate These five elements of literacy will provide beginning support in ensuring that the implementation of technology in literacy is done in a way that engages, educates, empowers, embeds, and elevates literacy instruction for your learners.

Engage for Effective Literacy Instruction

Think about ways engage you may engage students. Some survey questions for you to consider may include:

1. *What is considered popular with your students?*
2. *What movie, song, color, phrase, and or activity is gaining the interest of your students?*
3. *Have you used technology tools such as Google Jamboard™, Padlet™, or a digital poll to identify student interest?*

There are many technology tools that can serve as a great resource to use with your students to capture student interest and preference for learning. You may also consider using TikTok™ or other social media trends, books, commercials, current events, and gaming applications. The benefits of the survey questions support identifying interests and establishing rapport with students. This information can also guide planning out engaging lessons, assignments, projects, and connections to the real-world. Engaging, rigorous hands-on, and grade-level appropriate lessons can derive from collecting this information. Educators are challenged to think outside the box and tap into creative ways to teach new things from a different perspective to support student buy-in, engagement, and instruction. We must understand what students need and what commonalities students use to establish and maintain understanding and development.

As educators, we should also consider topics of interest when introducing a new concept or skill to a class. Identify what will keep them engaged and anticipate and plan for possible misconceptions. When addressing misconceptions, utilize real-world experiences to effectively plan for misconceptions, scaffolds, and various modalities of instruction to support students' understanding. After using data from your interest inventory to hook and engage your students in the lesson, implement differentiation and student application of new learning to build capacity. Remember that creativity will be essential! So practice providing students the autonomy to submit responses in various ways with alignment to the lesson objective.

Engage with Gamification

Are you having a hard time getting students to understand the concept? Do you provide opportunities for students to lead learning? Is there a giant pile of papers on your desk at the end of the day? Do your students understand the lesson that is taught? How do you know? Have you considered gamifying your classroom?

There are various ways to turn a classroom challenge into an opportunity for a game for students to work towards success. Students can sometimes better understand concepts, consequences, and rewards once the challenge has been gamified. Gamifying instruction increases student engagement and outcomes. When introducing new vocabulary and literacy concepts, consider ways to integrate games. Gamifying instruction creates a positive classroom environment and supports the development of goal setting, fun, and accountability. Students have the opportunity to practice problem-solving and research solutions. This practice is beneficial for students having challenges understanding concepts. Increase technology and opportunities for students to partner in pairs, groups, or teams. Create a game to address the misconceptions and students' understanding. Students learn from their peers. Allow them the opportunity to research, discuss and create to solidify and practice the application of learning. Think about using online platforms to support students working collaboratively while playing games. Kahoot!™ [1] and Quizziz™ provide platforms to support learning in a fun way.

Educate

Accountability and leadership are essential for both students and teachers. Instead of providing students with information, constantly think of opportunities for them to incorporate movement, song, and creative expression throughout instruction. With support, encouragement, and practice, students can obtain any goal set before them at any age.

1 Kahoot! and the K! logo are trademarks of Kahoot! AS.

Students that select the correct textbook or locate the story for the read-aloud have already established a "win." We want students to have as many small "wins" as possible to ensure they are engaged in instruction. These "wins" ensure they are actively engaged while developing their leadership and problem-solving skills. Understanding the importance of these opportunities will ensure students succeed during and after the lesson because they are bought in.

Research shows that implementing effective literacy instruction increases the development of students' comprehension, fluency, and vocabulary. The use of digital resources such as Brain Pop®, Brainpop Jr. ®, Flocabulary™, and Quizzz™ assist in educating students about literacy. YouTube*™* provides many videos related to literacy that are great for introducing new concepts and vocabulary. When showing any video, it is essential to preview it before presenting it to the class and plan for stopping points to check for student understanding.

Educators should ensure lessons delivered are engaging and rigorous. What systems, visuals, and activities have you secured to ensure your students are excited about learning and applying what they've learned? Have you created and scaffolded questions for students on all tier levels to ensure equity within instruction? Remember the importance of creating a "win" for every student and address misconceptions using real-world examples that students can connect to. Implementing videos from the sites listed above will support the students in making connections to learning. Use technology to support visual and verbal learning. When planning for instruction, always review curriculum and research opportunities to incorporate technology to engage, educate and empower students' comprehension.

Empower

Think about guiding questions that are connected to student interest and extending thinking. Don't forget to scaffold to ensure every child feels successful throughout the learning. We have to understand and continuously remember that one size does not fit all. So, as teachers, our goal is to constantly reflect, assess, survey, and make modifications to ensure our engaging lessons meet every child's needs. After identifying student interest and need, creatively plan ways to implement effective, rigorous, and engaging instruction.

Engagement can also consist of active participation. For example, project the book using a smart board or projector during a read-aloud. This is an opportunity for students to use journals to take notes or bring the text or book to follow along with the story being read. Students can use their table of contents to identify the story's location. During this time, encourage and plan opportunities to collaborate with classmates to ensure every student successfully navigated to the correct page. Once students have successfully located the story, ask them to give a non-verbal gesture to check for progress. Some examples of nonverbal gestures can consist of; thumbs up, thumbs down, silly facial expressions or thinking facial expressions, dance, or a gesture related to a character or the central idea of the story they will read. You can also incorporate a verbal song or response related to their interest. Include these engaging checks for understanding throughout your read-aloud to keep students engaged. Be sure to plan out these opportunities and provide explicit instruction to ensure you are successfully maximizing instructional time.

Plan opportunities to reward and encourage student success. Use Class Dojo™ Big Idea Videos to empower students learning using Social and Emotional lessons that focus on growth mindset, perseverance, empathy, gratitude, mindfulness, moods, attitudes, significant challenges, respect, and positive thinking.

Embed

Once students have demonstrated that they are equipped with their materials on the correct page and have utilized one of the non-verbal or verbal signals established for the lesson, the implementation of the read-aloud can commence. Projecting stories on a large or visible screen for students to follow along in their book is an effective and engaging visual for students to reference throughout the lesson. When you begin the read-aloud, ask questions throughout the story, such as what is the title of our book? What is the page number our story is located on? Describe what you see in this picture? Based on this picture, what can we infer about the story we are about to read? Suppose you notice students are disengaged or having a challenging morning. In that case, you might try saying something silly like, "oh my goodness, this picture looks like a dog and a cat are arguing over a cheese sandwich. I can tell this will be an interesting story". In reality, the story displays a picture of a family preparing for a trip. This is another strategy used to hook students. Often students respond boldly and ask for clarity because it is evident that the example provided is not visible on the book cover displayed. This is an opportunity to pose questions such as: How do you know? Describe what you see. Prompt students to explain and defend their descriptions and extend their thinking on and retrieve predictions of the text. Students can use voice-to-text tools to verbalize and record their thinking and responses. Access prior knowledge using Google Jamboard™ or Padlet™.

Allow students the opportunity to use connections to events they may have experienced to support buy-in and increase engagement. Ask them to add their thoughts, pictures, and written responses to these online platforms. Ensure students have the opportunity To apply information that they have learned digitally and verbally. A great recess to support collaboration and verbal response is Microsoft Flip™. Flip can creatively support verbal responses. Students can respond to each other's videos in the comments section. There are many ways to use Microsoft Flip™ to introduce, practice, and apply vocabulary and submit written responses to capture exit tickets for standards taught for the lesson. Always consider opportunities for student discourse in whole group instruction, small group instruction, and independently. Collaborate, respond, practice, and apply learning to identify misconceptions and extend learning.

Elevate

How are you using technology to partner with parents to support student development? Class Dojo™ is a resource that can be used to bring parents into your classroom virtually. Class Dojo™ is used to take photos, reward students using a points system, and notify parents of areas of development. This platform also supports ongoing communication with parents and guardians to ensure alignment and support are communicated regularly. Elevate your instruction by using this platform or other online platforms to increase family communication and exposure to instruction. Challenge and encourage students continuously. Literacy can be challenging, but we want to ensure we engage and encourage our students to master each standard and skill successfully. As educators, our overall goal should always be to engage, educate, empower, embed, and elevate.

5 TIPS TO SUPPORT IMPLEMENTATION

Engage: Game On - Gamify your Class!
Engage students in a way that allows them to connect, respond, ask questions and have fun while learning. If they are bored, they are not learning.

Educate
Educate students by using visual, auditory, kinesthetic, and tactical styles of learning that are relevant to the world they know. Make everything fun.

Empower
Empower students by providing affirmations, a positive classroom and culture, a growth mindset, and the skill set to support them in solving any problem they experience. Be their

Embed: Research for 500 Please!
Embed knowledge in students by providing opportunities to collaborate, respond, practice, clarify and extend literacy instruction. Researching skills will take them far.

Elevate: Use Your Village
Elevate your partnership with families and communities to reinforce and increase student learning. It takes a village to create a leader.

ABOUT ERIN CARROLL

Erin Carroll has served as an educator for twelve years. Her background includes roles as a youth counselor, social services case manager, classroom teacher, department chair, literacy specialist teacher leader, and literacy teacher development specialist in private and public schools in urban areas. In her latest role, Erin Carroll serves as an Academic Program Manager for Teacher Career Development Literacy Teacher leaders for the largest district in Texas. Her professional development and support of teachers and teacher leaders have resulted in change and improvement of effective literacy instruction and coaching. She specializes in support of urban area campuses in grades K-12. She has contributed to increased reading levels and student comprehension within multiple districts based on inquiry-based instruction, digital integration, data-driven instruction, literacy routines, and student engagement. She believes every child can learn, and within an engaging, rigorous, project-based environment, every child can exceed all grade-level expectations. She believes learning can and should be fun, and all educators should ensure students connect to learning and extend their thinking in a way they can relate to.

"Innovation continues to fast track our daily experiences. It is essential to provide opportunies to engage with innovations of our time through technology, literacy, and STEM experiences.

- Jocelyn "Dr. Mac" McDonald, Ph.D.

PART 4
ADVANCING INSTRUCTION WITH INNOVATION

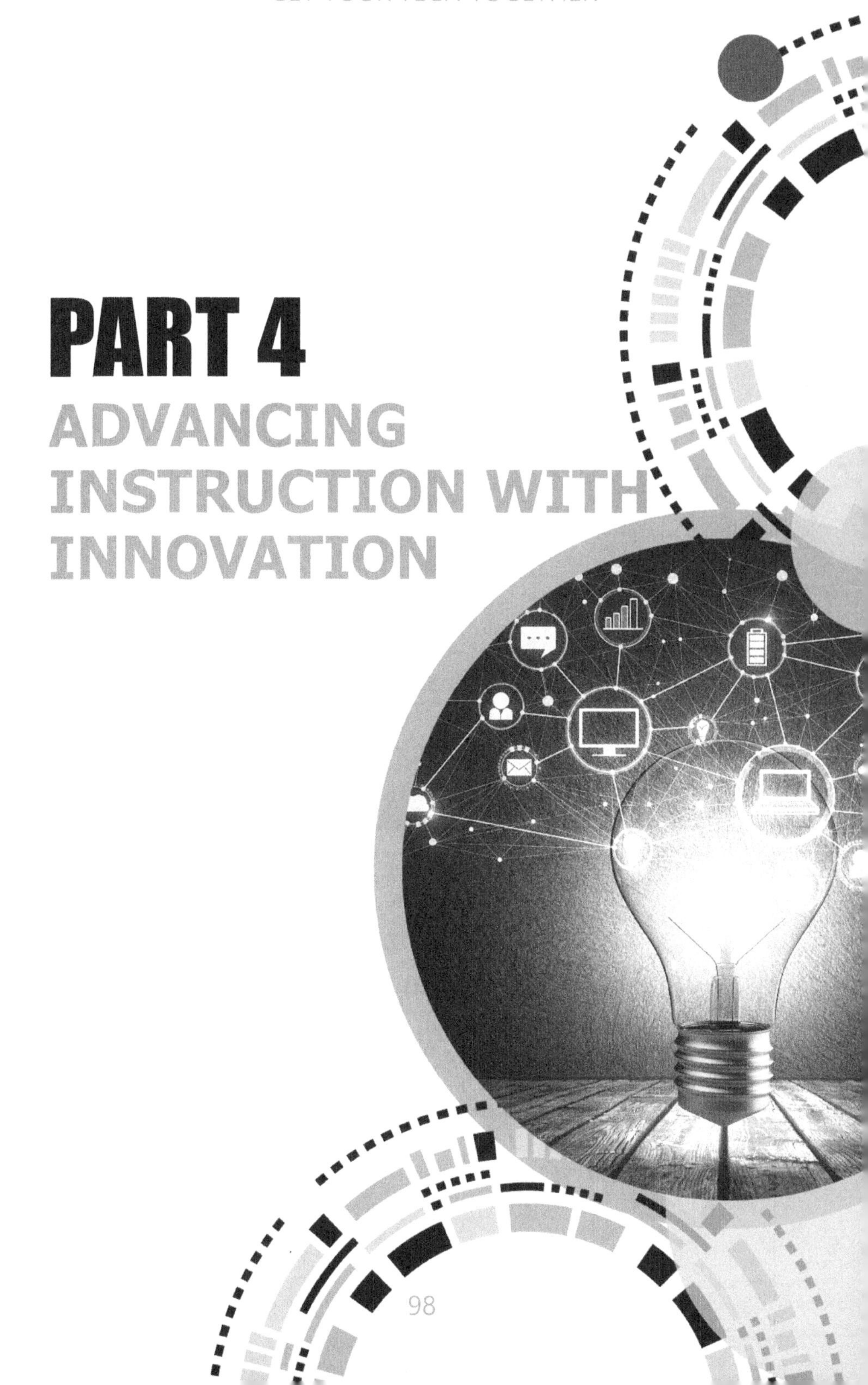

SHIRLEY POSEY

"The key to revolutionizing education will be driven through the advances of technology."

- Shirley Posey, M.S.

Chapter 9

If Technology Advances Why Shouldn't Our Educational Practices?

by: Shirley Posey, M.S.

As a teenage girl, I recall being ecstatic about my Nana getting a mobile phone in her Victorian Crown car. Then in 2007, Apple made a meteorite impact worldwide with the iPhone. This smartphone consisted of a metal phone, making the physical hardware buttons and stylus for a screen-based interface obsolete. The iPhone has adapted and continues to change to meet the needs of current society. Super Retina XDR ProMotion technology allows up to 25 percent brighter outdoors, more efficient battery life, and solutions to challenges with previous models. As educators and leaders, we must ask ourselves if our educational practices and delivery are evolving.

> "The world has changed far more in the past 100 years than in any other century in history. The reason is not political or economic but technological" -Stephen Hawking

Technology plays a significant role in 21st-century skills needed to advance our world, drive the economy, and solve real-world problems. The importance of integrating technology into our learning spaces allows exposure, experience, cognitive development, student engagement, and academic success. Understanding that technology plus imagination multiplied by creativity equals discovery was the formula that I cracked and caused me to become a transformational educator and visionary in STEM education. I realized that without our pedagogies evolving integration of technology, we educators would not be able to properly equip our scholars to be change agents or a part of the equation with solving global problems.

In 2020, during the pandemic, this theory was tested in my professional context. How do I transform Zoom into an academic platform that allows students to still develop 21st-century skills, perform proficiently on standardized testing, and ignite intrinsic motivation? The solution was technology mixed in with creativity to produce an innovative and immersive virtual learning space! But, it didn't just stop in the Zoom world. These critical concepts, aka the STEM formula I developed, allowed me to elevate my learning space through technology and produce positive academic outcomes.

The following technology tips will support you and guide your students through experiential learning, increase engagement, enhance academic performance, meet the learning needs of diverse student populations, and be culturally inclusive.

Revolutionize Education with VR headsets

A clinical case study by Beijing Bluefocus E-Commerce Co., Ltd and Beijing iBokan Wisdom Mobile Internet Technology Training Institution (2016) [1] investigated The Impact of VR on Academic Performance. Subjects were astrophysics high schools in Beijing that collected data comparing students'

1 Beijing Bluefocus E-Commerce Co., Ltd. and Beijing iBokan Wisdom Mobile Internet Technology Training Institutions, 2016. Research Study Suggests VR Can Have A Huge Impact In The Classroom.

Photo taken and owned by Shirley Posey.

exams into two groups: virtual reality instructional teaching (VRIT) and traditional teaching. The VRIT group received an astrophysics lesson instruction using the virtual reality headset, and another group received instruction through traditional practices. The average score of the VRIT group is 93%, and traditional teaching is 73%. On average, the VRIT group has registered a 27.4% growth in terms of score, indicating the great advantage of VR in the teaching of astrophysics. The study revealed that by presenting to students the abstract contents in the textbook in a three-dimensional way, VR takes advantage of helping students better understand and acquire the knowledge. VR-based teaching is vivid and interactive. It helps students get better test scores by enhancing their understanding and memorizing knowledge. The test results were drastically different. With the new wave and excitement for VR headsets and oculars, it can be an easy distraction to think that headsets are only for entertainment. However, these headsets are designed to create immersive learning experiences and environments. Imagine being able to manipulate an atom or create ionic charges virtually instead of just staring at a 2D picture or manipulating it with pencil and paper. VR headsets would also allow students to understand the importance and significance of learning foundational content through application.

How did I apply this technology in my class? The students were assigned to build a wheelchair out of PVC pipes for underprivileged people in Kenya. However, before designing their wheelchair prototype, the students used the VR headset to immerse themselves in the streets of Kenya. This experience allowed them to understand that they could not just design a wheelchair with the design that we use in the United States because the terrains are different. Just by allowing the students to immerse themselves in the rocky and rural roads of Kenya, students began to create a prototype that was innovative to the traditional wheelchair. Within our STEM department, we also used VR headsets for our students to design drones and understand photosynthesis and solar energy communities.

VR headsets create immersive learning environments that are for diverse learners. These environmental learning spaces breathe life into the textbook, and augmented reality (AR) allows those spaces to become manipulative. Immersive learning experiences enable facilitators to cater to diverse learners. Students should not be viewed as "passive learners" but as active participants in their learning process producing advocates and global change agents.

Allow Students to Connect Problems in heir Community to Solve with Technology

The Brain Targeted Teaching Model V discusses the importance of taking skills and knowledge taught in class to solve problems in their community, thus connecting learners to the classroom in a real-world context. Dr. Ayesha Imani, founder of Sankofa Freedom Academy, dissertation 'From SBA to HEKA: An Examination of the Community Service-learning Practices' discusses a culturally inclusive service model that allows students to connect their studies beyond the scope of the classroom. [2] This framework provoked

2 Imani, Ayesha, 2005. From SBA to HEKA: An Examination Of The Community Service Learning Practices in Three African Centered Urban Schools. Temple University ProQuest Dissertations Publishing. 3178790.

scholars to promote advocacy and change in their community. So, how do cultural inclusivity and STEM skills bring effective change? Through innovative technology.

For instance, my scholars applied the Engineering Design Processing (EDP) and Newton's Third Law of Motion to design drones with cameras, and students controlled drones with an app on their phones. Then the purpose of the meat was integrated into the project--a question. How does this solve a problem or address a community problem? Within our student population, the first problem was --Gun violence. Posed ideas that the drones could be used in high-crime neighborhoods for surveillance and perhaps even quickly alert the authorities on shooting in the area. Under the leadership of Dr. Imani, the advocacy and student voice was led in the "enough is enough" campaign meeting with political leaders to bring about effective change in gun violence in the community. A beautiful merge of social justice and technology working together to produce authentic learning.

Imagination is the Secret Sauce to Technology

Within the neocortex of the thalamus reside areas of neurons that create abstract and creative thoughts. Okay, let me break this down. In the center of your brain, there is a network of brain cells, aka neurons, that communicate with each other through electrical and chemical signals influenced by external and internal stimuli. In this case of imagination, the neural network in the intersection of the brain called the thalamus is triggered by an outer world (environmental) stimulus that causes it to think abstract or creative ideas-- typically resulting in innovation or invention. However, our educational spaces are plagued with standardized testing that we omit developing the concepts that govern our economy and world-- imagination. Why not create a double edge sword in your learning space, providing ideas, standards, and pedagogy with an imagination-building lesson to develop that skill? How do we achieve this by incorporating technology into the learning process? One of the ways I reached this was through Microsoft Flip™ (formerly Flipgrid). Microsoft Flip™ is more than just a way to respond to questions posed by facilitators; but allows students to creatively express their thoughts and learn from each other. With the special features, it also teaches students to learn 21st-century editing skills. For example, I incorporated this into our Forensic Science laboratory, where students had to creatively use evidence to explain who committed the crime and discuss the importance of accurately collecting DNA evidence. Students were able to creatively deliver, educate, and apply lessons learned in genetics lessons. The academic performance and engagement were astounding.

Even as the times change, so must our educational practices. I would like to challenge you to use the neural networks in your thalamus to dig deep and incorporate technology into your lessons to revolutionize and further develop the learning process of our scholars.

5 TIPS TO SUPPORT IMPLEMENTATION

Tap into your 5-year-old creativity when you could turn a water bottle into a sword.
Harvard School of Business defines a creative genius that has mastered combining divergent and convergent thinking skills to continually produce innovative and novel ideas. Use this creative genius to implement technology into your lesson.

Be the real-life Miss Frizzle
Miss Frizzle is a cartoon science teacher who goes with her students on incredible field trips using a magical school bus. This character adopts an inquiry-based technique for teaching various science concepts. For instance, if she's teaching about the heart, the school bus will transform into a red blood cell, and the students will have to figure out a solution to solve the problem once they enter the human body as a cell. Use this concept to create interactive and immersive experiences and spaces within your classroom using projector screens or VR headsets.

Allow students to connect problems in their community to solve with technology
Don't teach about it, be about it. When students can relate to the material presented, it causes them to develop creativity, critical thinking, collaboration, and communication. These skills can allow students to place a STEM lens on solving innovative problems within their community locally

Revolutionize Education with Technology
It's okay if your peers, students, and parents know that you don't know everything. We learn best from each other and should not be afraid to show our vulnerability and weaknesses.

Technology is not a trend, but a tool
Virtual Reality headsets, Merge Cubes (project holograms), and 3D printers sound fantastic, and some can be distracted by the entertaining features. But, technology can evaluate and maximize the

ABOUT SHIRLEY POSEY

Shirley Posey has been an educator and leader in education for over twelve years and is known for her expertise in evolving learners for the future. She has used her expertise as Director of STEM at Imhotep Institute Charter High School in Philadelphia, Adjunct Professor at Eastern University, and Johns Hopkins University CTY Instructor. Shirley has applied the investigative and innovative essence of STEM, developing forward-thinking lessons, immersive STEM spaces, and progressive STEM curricula creating global problem solvers. Shirley earned her B.S. in Biology from Clark Atlanta University and her Masters in Medical Science from Hampton University and is currently a doctoral student at Johns Hopkins University. She co-founded NeuroAdvanced Academics, Inc., an educational firm using brain-based practices to enhance academic learning. Shirley consults and facilitates Professional Development for educators around the country on creating innovative STEM practices. She is the 2020 Philadelphia Eagles and Axalta Educator of the Year for the tristate for contributions to STEM education. She has been a TEDx Talk, speaker at the DANA Foundation Brain Awareness Week, Philadelphia School District Parent Symposium, Nashville SEL Conference, 28 days & Beyond with Villanova University, and features in the Philadelphia Tribune, ABC Philadelphia, and NBC Philadelphia, and Liberty County Tribune. Shirley is the loving mother of the brilliant and vibrant Elizabeth Elle.

This chapter is dedicated to my seven students whose lives were taken early. Your memory will forever be sketched in my heart.

CONNECT WITH SHIRLEY POSEY

@shirleyelleposey
@shirleyelleposey
shirleyelleposey.com
info@shirleyelleposey.com

DR. KRISTINA TATIOSSIAN

"Learning from books is good, learning from experience is great, learning from other people's experiences is extraordinary!"

- Dr. Kristina Tatiossian

Chapter 10

Closing the Gap Between STEAM Innovation and Education with Next-Generation College and Career Preparation

by: Dr. Kristina Tatiossian

What is the purpose of our education system? Is it to prepare our kids for a successful life, including getting paying jobs to sustain that life? To become lifelong learners with critical thinking, communication, and collaboration skills? To effectively serve our communities?

You wouldn't be alone in answering yes to the above questions – in fact, nearly all educators we interviewed had some variation of this same response.

Our education system is designed to enable all children to achieve their highest potential, serve effectively as citizens in our society, and successfully compete in a changing global marketplace.[1] And in the last decade, Science, Technology, Engineering, Art, and Mathematics (STEAM) education has grown increasingly critical. In this age of innovation, young people need STEAM skills to succeed in school and beyond because, as we know, today's students are tomorrow's leaders. And as the world of science and technology continues to expand, we must develop our students to

1 UNESCO-IBE. "World Data on Education." UNESCO, 2006, http://www.ibe.unesco.org/sites/default/files/United_States_of_America.pdf. Retrieved Aug 12, 2022

become the future workforce that thinks critically, solves complex problems, and addresses our most pressing issues with logic and creativity.

The challenge, though, is that STEAM fields are constantly evolving, making it difficult for education to keep pace. Nevertheless, continually updating education to meet the demands of modern life is essential. In fact, failing to update curricula makes students less likely to go to and graduate from college, invent and secure patents, and get a higher degree.[2] This phenomenon is known as the gap between innovation and education. And as you can imagine, the wider the gap becomes, the less likely students are to fulfill the purpose of our education system.

As a result, we must ensure students have access to frontier knowledge applicable in the modern world to prepare and motivate them to step into their future properly.

How to Improve Student Outcomes by Minimizing the Gap between STEAM innovation and Education

As a scientist, I work at the bleeding edge of innovation. I recognize that STEAM fields evolve very quickly, and in my line of work, information can become outdated in as little as 6 months. While educational curriculum and programs will not be able to keep pace 1:1 with STEAM innovations, educators can still use specific strategies to minimize the gap between innovation and education at their schools and improve student outcomes. These strategies include incorporating innovative curricula, ensuring students have access to hands-on learning experiences, modern professional development, and finally, a key but often overlooked engagement model that seeks to promote college and career readiness by directly connecting

2 Biasi, Barbara, and Song Ma. "The Education-Innovation Gap." National Bureau of Economic Research, May 2022. Retrieved Aug 10, 2022

students and teachers with STEAM industry professionals (figure 1).

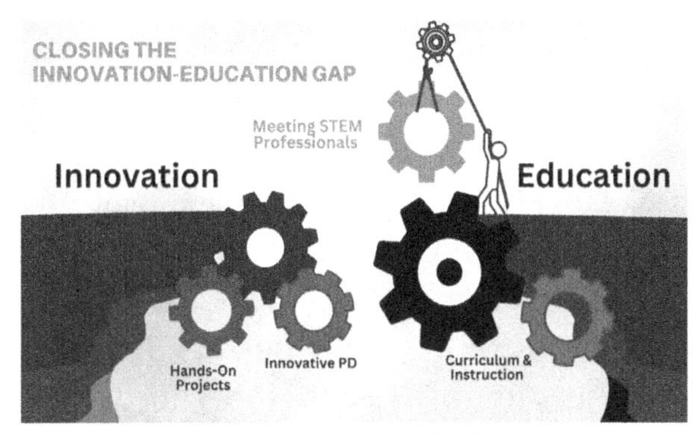

Figure 1. Closing the Innovation-Education Gap

In fact, when students meet scientists, they...

- Learn cutting-edge STEAM from subject matter experts
- Improve awareness of STEAM careers
- Overcome biases about who they believe can work in STEAM
- Relate their classes to the real world.

Additionally, students hear real people's stories, make connections, get inspired, start to feel the human side of science, and so much more.

Connecting your students to scientists to close the gap between innovation and education can be accomplished in three possible ways. The first option is to host career events. Career events often feature multiple professionals, occur on specific dates/times, and are centered around short interactions with scientific professionals. This is an excellent opportunity to expose students to many career tracts within a brief period. However, this structure also has specific shortcomings. For example, students have limited time to speak with and understand each career opportunity. Additionally, career events are often complex to host and coordinate. Finally, career events disfavor more introverted personalities that can be easily overwhelmed, leading to mental exhaustion, decision paralysis, and apathy in students. Therefore, implementing career educational events is challenging for the host and may not be optimal for the student experience.

The second option is to partner with a neighboring research university and create a program where your students can meet their scientists-in-training continually throughout the school year. Research university partnerships are a fantastic way to get your students more involved and interested in STEAM. During my Ph.D., the high school down the block partnered with my university and sent our laboratory interns to be trained in CRISPR gene editing. If every school in the US had this kind of access, the gap between STEAM innovation and education would not be so wide. However, these programs are scarce because they are difficult to develop from scratch! Not only do you have to find a research university in your region willing to partner, but you will also need to determine the structure of the program, including how, when, and for how long students will meet with researchers, and which researchers are willing to be included in the program. Furthermore, such a strategy comes with inherent limitations, including: (1) the researchers you have access to will all be graduate students or post-doctoral scholars (post-docs), lacking completely STEAM professionals who work in industry, and (2) the researchers will not be trained for student engagement, and finally (3) the true diversity in science cannot be showcased because you're relying on a relatively small pool from a regional research community.

The third and final option is to leverage virtual spaces to invite scientists to your classrooms. While 100% virtual education is not optimal for student outcomes, and we've all experienced "zoom fatigue," we should not eliminate virtual opportunities altogether! When scientists are continually brought to your classrooms through virtual spaces throughout the school year, you expose students to a more diverse pool of STEAM researchers without geographic limitations. One way to find scientists willing to speak with your students is to leverage Twitter and LinkedIn specifically. You can leverage this network to find, contact, and ask scientists to visit students virtually. Though I'm confident many of the invitations will be accepted, a response may not be guaranteed. The individual cannot be pre-screened and will likely not be trained for student engagement. Additionally, all the lesson planning and preparation for the scientist visitor are still left to the classroom teacher.

Alternatively, you can rely on 3rd party matchmaking programs like STEAMmates.[3] STEAMmates is an easy-to-use platform developed by scientists that allow educators to invite diverse STEAM professionals to classrooms virtually. STEAMmates is fully turnkey to overcome many of the challenges listed above. From simple scheduling to pre-built, social-emotional aligned lesson plans, classroom teachers are prepped and ready to go from A to Z. It's as easy as (1) the educator requests the scientist or specific meeting time, (2) the match is made, and (3) the students meet the scientists!

"I have seen kids that are low achievers, that have low motivation, don't do very much···some of the few times that I've actually seen them engaged and excited, and doing something, [are] when STEAMmates were there." - Dr. Wilson, Ed.D., Principal, Pennsylvania

Regardless of the methods, the purpose of bringing scientists to students remains the same. Closing the gap between STEAM innovation and education is essential to ensure students fulfill the purpose of our education system. Teaching innovative and modern methods is challenging. However, based on proven pedagogical practices, we know innovative curriculum, hands-on projects, and access to career preparatory programming are the backbone that ultimately improves student test scores, graduation rates,

3 CRISPR Classroom. "Steammates." https://www.crisprclassroom.org/steammates

5 TIPS TO SUPPORT IMPLEMENTATION

Leverage LinkedIn and Twitter to find scientists

Scientists are very active on both LinkedIn and Twitter. Though it may feel intimidating, many are open to volunteering to visit classrooms virtually. Go ahead and drop them a message!

College and career events are key for students to understand the "why" of learning

Universities, particularly research universities, often make partnerships with neighboring schools! Leverage your neighborhood to give your students access to the frontier technologies being developed.

Make Partnerships with Neighboring schools

Discover creative ways to provide a welcoming learning atmosphere for your students. Allow students to set personal goals for success and self-reflect on their learning.

Use paid programs whose purpose is to connect your students to scientists

Programs from CRISPR Classroom or HealthWorks Academies, two institutions dedicated to sharing the lives and experiences of scientists, connect classrooms to biotechnology professionals. Leverage them for easy access to scientists.

Get Buy-In from administration

Regardless of how you want to improve your students' learning experiences, getting buy-in from the administration will be key. Luckily, introducing students to scientists always has a nice ring to it.

ABOUT DR. KRISTINA TATIOSSIAN

Kristina Tatiossian, Ph.D. ("Dr. Kris") is a scientist, entrepreneur, and consultant passionate about enabling the next generation of students to pursue STEAM careers. She has worked in top biotechnology companies developing novel therapeutics for the treatment of HIV, sickle cell disease, and more. Dr. Kris is the founder of CRISPR Classroom, an education technology company closing the gap between STEAM innovation and education. CRISPR Classroom specializes in science storytelling, thereby granting students, educators, and parents access to learn frontier scientific knowledge. Additionally, Dr. Kris works closely with various for-profit and non-profit STEAM organizations in her role as a digital marketing consultant and social media growth hacker.

VISIONARY AUTHOR

" Leverage technology as a vehicle to create meaningful learning experiences that mirror the digital world our students are brought up in and prepare them for the future they will contribute to. Lead. Inspire. Develop."

- Jocelyn "Dr. Mac" McDonald, Ph.D.

ABOUT JOCELYN MCDONALD, PH.D.

Dr. Jocelyn McDonald is an Educational Technologist, Podcaster, Blogger, and founder of #TechItUp, LLC. #TechItUp, LLC is a professional learning and technology solutions company dedicated to supporting educators and educational leaders with the implementation of technology to create meaningful student learning experiences. She has trained and developed thousands of teachers, school administrators, instructional specialists, and various educators on effective practices to integrate technology, instructional design, efficient and sustainable systems, as well as change management and strategic planning. She is a proud advocate for social change in education and is dedicated to making a positive impact. She holds a Bachelor's in Chemistry, a Master's in Curriculum and Instruction in Instructional Technology, and a Ph.D. in Educational Technology. With 16+ years in education, she is passionate about supporting the transformation of education through technology for equitable access to 21st-century education while also serving as a Director of Digital Learning. She makes strong efforts to use her influence to impact achievement in schools by building relationships with various stakeholders and looks forward to closing equity and diversity gaps in education.

Made in the USA
Middletown, DE
10 October 2022